Food and Human Evolution

Food and Human Evolution

How Ancestral Diets Shaped Our Minds and Bodies

Berman D. Hudson, Ph.D.

Algora Publishing
New York

Library of Congress Cataloging-in-Publication Data

Names: Hudson, Berman D., author.
Title: Food and human evolution : how ancestral diets shaped our minds and bodies / Berman D. Hudson.
Description: New York : Algora Publishing, 2021. | Includes bibliographical references and index. | Summary: "Food has played a major role in human evolution. The fact that we stand upright, that we can talk, that we have big brains; even traits such as altruism and a sense of fairness — all of these can be attributed largely to the kinds of food our ancestors ate and how they acquired it. We now face a modern food-related crisis. This book describes how the rise of industrial food production during the 20th century unleashed an epidemic of metabolic disease that now threatens the very future of our species"— Provided by publisher.
Identifiers: LCCN 2021025058 (print) | LCCN 2021025059 (ebook) | ISBN 9781628944679 (trade paperback) | ISBN 9781628944693 (pdf)
Subjects: LCSH: Prehistoric peoples—Food. | Food habits—History. | Human evolution. | Human behavior—Nutritional aspects. | Diet in disease.
Classification: LCC GN799.F6 H84 2021 (print) | LCC GN799.F6 (ebook) | DDC 394.1/209--dc23
LC record available at https://lccn.loc.gov/2021025058
LC ebook record available at https://lccn.loc.gov/2021025059

Printed in the United States

I dedicate this book to the people who have given me the greatest happiness — those who call me Dad or Grandpa. A special thanks to my granddaughter, Zoe Curtis, for her meticulous and insightful editing during the writing of this manuscript.

Table of Contents

Chapter 1. A Modern Dilemma

Sometimes we can be oblivious to the most profound changes taking place around us. This was true in my case. For many years I was blind to the fact that Americans were becoming obese at an alarming rate. It took a disconcerting event to open my eyes. This occurred some years ago during a vacation in the Black Hills of South Dakota. At a popular tourist site, I shared a lunch table with an English family that was taking a bus tour through the American west. During our conversation I asked them what they thought was most striking about America, expecting them to describe some famous landmark or point of historic interest. I was not prepared for the answer. The father looked around carefully to make sure no one else was within hearing distance. Then he leaned forward and whispered with a distinct tone of embarrassment, "There are so many fat people everywhere in America, and they are so huge!"

"Do you mean there are no fat people in England?" I retorted a little sharply — in fact, a little defensively.

"Oh, of course there are, but there are not nearly so many. And there are fewer truly obese people like one sees everywhere over here in the States." His wife and two daughters nodded in silent agreement. The two little girls were wide-eyed, apparently amazed at the sheer beefiness of their American cousins.

As a result of this chance meeting, I began to look at myself and my fellow Americans with heightened awareness and to notice the fattening of our country. About a year after talking with the English tourists, I stopped in a small town in eastern Oregon. One of the local restaurants had group pictures of the Future Farmers of America club from the local high school going back to the 1940s mounted on the wall. I spent a fascinating hour or so carefully examining these pictures. It was amazing to see how teenagers had changed over time. Throughout the 1940s and 1950s there were no truly obese kids in the pictures and very few one would classify as overweight, but starting in the 1960s things began to change. As the years went by, a few of the kids became a little heavy, and then a few became obese. Finally, by the mid to late 1980s overweight and obese teenagers dominated most group pictures. Truly lean or skinny kids had become a rarity.

This experience intrigued me. As a result, I later examined the yearbooks of several American high schools in different cities. The same pattern was repeated in every case. Reviewing old issues of American magazines, especially picture magazines such as *Look* and *Life*, provided further confirmation. Group and crowd pictures from the 1940s and 1950s depicted an America much leaner than today.

America is not alone; obesity and its attendant ills are rapidly becoming a global problem. The United Nations issued an alarming report in 2004 titled *Fighting Hunger Today Could Help Prevent Obesity Tomorrow*. This report warned that millions of malnourished children in poor countries could be ticking obesity time bombs. As their countries become more affluent, many of these children will have access to large amounts of high-fat, high-calorie foods for the first time in their lives. The resultant weight gains could be sudden and massive. Poor countries still struggling to cope with infectious diseases could at the same time face epidemics of obesity, diabetes, and heart disease. Obesity, once perceived as a singularly American or western affliction, is fast becoming a global epidemic. Rich or poor, it appears that few nations will be spared.

A Modern Phenomenon

Until very recently, in evolutionary terms, nearly all humans remained lean throughout life. Obesity entered the human race slowly at first, beginning to affect the wealthy elite soon after the advent of agriculture. Obesity was known to the early Greeks and Romans, but very few people were affected. The English and the Western Europeans began to recognize it as a worrisome problem among the wealthy about 300 years ago. It became widespread only during the last century, and did not begin to reach pandemic proportions until the last few decades.

The National Portrait Gallery in London's Trafalgar Square provides a striking pictorial account of how the obese phenotype first appeared among the English nobility. The gallery contains portraits of all of the

English monarchs from the 14th century on. None of them appears to be overweight until near the end of the 1400s, when Henry VIII has a slight double chin. True obesity does not make an appearance until the last half of the 1600s, at which time Charles II is portrayed as being visibly obese, with a huge protruding abdomen. By the end of the 17th century, every member of the nobility, lord or lady, has become overweight or obese (Galton 1976).

What does the term obese mean? How do doctors decide if someone is obese or not? In most of the world, medical professionals use *body mass index* to judge whether someone is normal weight, overweight, or obese. Body mass index is based on the *Quetelet Index*, developed by Belgian mathematician Adolphe Quetelet in the mid-1800s. In 1972, American scientist Ancel Keys gave the Quetelet Index a new name, calling it the body mass index, now commonly referred to as BMI.

To calculate BMI, you take a person's weight in kilograms (w) and divide it by his or her height in meters squared (h^2). To calculate BMI using Imperial units, first divide a person's weight in pounds by his height in inches squared. Then multiply the answer by 703. Consider a man who is 5 feet, 9 inches tall and weighs 271 pounds. To calculate his BMI, you first multiply his height in inches (69) by itself (69 x 69), which equals 4761. You then divide his weight by this number (271/4761), yielding 0.0569. Then multiply 0.0569 by 703, giving a BMI of 40. According to the National Institutes of Health, anyone with a BMI equal to or greater than 30 is considered obese. A BMI of 40 or greater means that you are severely obese, which raises a number of health concerns.

Disturbing Numbers

Recent statistics verify what most of us can readily see. According to the Centers for Disease Control, between 2000 and 2018, the percentage of American adults who are obese increased from 30.5 to 42.4 percent. During the same time interval, the percentage of severely obese adults nearly doubled, going from 4.7 to 9.2 percent. Nearly one in ten Americans is not just obese but severely obese. The trends are quite striking. Prior to 1900, it is estimated that only around five percent of adult Americans were obese. By 1950, this had increased to about ten percent and by 1970 to 15 percent. Table 1.1 shows this dismal progression. There is every indication that things could get worse; since 1970, the percentage of obese children in America between the ages of 6 and 19 has nearly tripled. The current rate of childhood obesity in America is about 20 percent, and shows no signs of stopping.

TABLE 1.1. PERCENTAGE OF OBESE AMERICAN ADULTS BY YEAR

Year	Percent obese
1900	5
1950	10
1970	15
2000	30
2018	42

There is another disturbing trend — the rapid increase in the number of people who are defined as morbidly obese. These are individuals with a BMI of more than 50. A man five feet, nine inches tall who weighs more than 350 pounds would fit this category. Such individuals often cannot fit into airline seats and find it hard or impossible to walk around the block. From 1986 to 2000, the prevalence of morbid obesity jumped from one person in 2,000 to one in 400. Many of us now living could see a time in which humans weighing 500 pounds will be common and those weighing nearly 1,000 pounds will not be considered remarkable.

In the spirit of *schadenfreude*, there are parts of the world worse off than America. According to the World Health Organization (WHO), the island nation of Nauru has an adult obesity rate of 61 percent. Rates are nearly as high among populations living on the other islands of the south Pacific — for example, Palau at 55.3 percent, the Marshall Islands at 52.9, and Samoa at 47.3. Interestingly, there are a few nations that have managed to develop modern economies but have thus far largely avoided the 20th century plague of obesity. These include Singapore at 6.1 percent obesity, South Korea at 4.7 percent, and Japan at 4.3 percent. The least affected nation on Earth is Vietnam, with an obesity rate of only 2.1 percent. It will be interesting to see what happens to these nations in the future.

Many researchers have come to believe that trying to stop the rise of obesity is futile; that it at best can be managed or controlled similar to the way diabetes and high blood pressure now are managed to some degree. As more and more of us become obese, drugs will be given to keep our weight within manageable limits, but any hope for a cure will be abandoned. Currently, if diabetes patients fail to take their medicine, their blood sugar will not stay under control. According to this scenario for the future, obese people will need to take their "obesity medicine" on schedule to keep their weight under control. Once patients begin taking the drugs, they will need to stay on them for life — or until complications and unforeseen reactions force a change.

Patients might have to take a number of medications at the same time, with each affecting the body in a different way. One might raise the metabolic rate. Another might lower fat absorption in the intestines. Still another might increase the production of leptin or other appetite controlling hormones. Making all of this work will require quite a balancing act, but it could be very profitable for drug companies and for physicians specializing in life-long weight management. Of course, if drugs fail to work, there is always surgery.

Some might think that expensive drugs and highly complicated surgical procedures reflect a high level of knowledge. But just the opposite is true. Such expensive and drastic treatments are testaments to our ignorance, our inability to understand what is happening to modern humans. Nearly all of the basic questions remain unanswered. No one can explain why the rates of obesity and diabetes began to surge during the 20th century and show no signs of stopping.

Diabetes

Every year more Americans are graduating from the ranks of the merely overweight to become obese, and then continuing on to become morbidly obese. Obesity itself is a health issue, but the metabolic disorders that so often accompany it are the real problem. Type 2 diabetes, a disabling and often fatal obesity-related disease, is of special concern. As of 2012, according to the Centers for Disease Control, at least one in eight Americans, or more than 40 million people, had been diagnosed with the disease. People with diabetes suffer from accelerated aging. They die at higher rates from heart disease, stroke, and kidney disease. Compared with other Americans, they are more likely to suffer from deteriorating eyesight and blindness, nerve damage, tooth decay, foot ulcers, and gangrene. In 2010, about 73,000 Americans underwent lower limb amputations as a direct result of diabetes (Gregg et al. 2014; Menke et al. 2015).

Diabetes is now the seventh leading cause of death in the United States. Think about this when you go to bed tonight. By the time you wake up tomorrow and live through another day, more than 5,000 Americans will have been diagnosed with diabetes. During those same 24 hours, more than 200 of those suffering from diabetes will undergo limb amputations, more than 100 will begin end-stage kidney treatment, and more than 50 will go blind. Treating diabetes and its complications already takes up more than

15 percent of our national health budget; it is frightening to think of what it could be in a few more generations (Fumento 1997, Grady 2008).

Doctors rarely use the words "adult onset diabetes" anymore, since the disease is now showing up even in pre-school age children. The number of diabetes sufferers worldwide is expected to double from 1998 to 2025, and the victims are becoming younger with each passing year. The Centers for Disease Control predict that one-third of American children born after 2000 will be diagnosed with diabetes at some time in their lives. The odds for African-American and Hispanic children are even worse — closer to one in two. Experts project that children diagnosed with diabetes at age ten could have their lives shortened by nearly 20 years (Kleinfeld 2006).

Diabetes was very rare before 1900, but became more common during the early part of the 20th century. Then, within only a few decades, the number of cases began to accelerate rapidly, and the disease now is reaching epidemic proportions. In 1890, Scottish physician Robert Saundby, former president of the Edinburgh Royal Medical Society, characterized diabetes as "one of those rarer diseases," noting that it was responsible for only about one in every 50,000 deaths. William Osler, whom many call the "father of modern medicine," reported that, of the 35,000 patients treated at Johns Hopkins Hospital during his first three years there, only ten had been diagnosed with diabetes (Osler 1892). But things soon would change. Elliott Joslin, a graduate of Harvard Medical School, documented the rise of diabetes during the early 1900s, and soon was using the term "epidemic" to refer its inexorable rise. He wrote in 1921,

> On the broad street of a certain peaceful New England village, there once stood three houses side by side. Into these three houses moved in succession four women and three men — heads of families — and of this number all but one subsequently succumbed to diabetes.

Fast forward to the present. In 2012, the Centers for Disease Control estimated that between 120 and 140 Americans out of every thousand now have diabetes, and a new case is diagnosed about every 15 seconds. Even experts cannot explain why so many Americans going about their normal lives and eating "normal" American foods are becoming so obese and sick. This is the surreal world in which we live — ever larger numbers of people told that they must have their legs amputated, their abdomens opened up, part of their stomach removed or closed off, and their intestines rearranged in order to, hopefully, avoid dying at an early age. Something has gone badly wrong in the modern world. One thing we know for sure; the human diet has changed profoundly during the last century; and those societies in which diets have changed the most are experiencing the highest levels of obesity, diabetes, and other metabolic diseases.

No species can radically alter the nature of its food supply without consequences; the human race has demonstrated this several times during its long history. The first time was when we learned to fashion stone tools and weapons and became big game hunters. Eating and sharing meat led to our big brains and the culture and technology that followed. Later we mastered fire and learned to cook. Cooking provided us with a wider variety of foods that were higher in calories and softer in texture. Eating this kind of food enabled our jaw muscles to become less massive, making way for even larger brains.

Our last great adaptation prior to modern times began to take place about 10,000 years ago when our species adopted agriculture and became grain eaters. Although agriculture fed more people on the same amount of land, they were forced to eat a diet of much lower quality. As a result, generations of humans became short, stunted, and disease-plagued. Perhaps half of humanity, mostly living in poorly-developed countries, has yet to recover from the effects of this.

We now are in the throes of another dietary revolution, and those earlier changes may pale in comparison to what is coming. Beginning about 100 years ago, people living in developed countries began to give up their traditional diets in favor of refined, pre-packaged, factory-made foods. This was a revolutionary shift — from the age of organic agriculture to what could be referred to as the age of industrial food. This already has begun to cause profound changes that do not bode well for our species, and more rapid changes are on the way. The industrial world is awash in vegetable oils, sugar, salt, high fructose corn syrup, and refined grains, and the consequences are becoming increasingly evident.

Thrifty Genes

In 1989, Neil Postman wrote an illuminating essay in the *Atlantic Monthly* titled "Learning by Story." His main thesis was that humans need stories to make sense of their existence. By story, he did not mean fiction. Instead, Postman was referring to the theories, legends, histories — even myths, that families, nations, religions and scientific disciplines create to bring order to their world. If we lack some kind of story to provide a framework for our observations and perceptions, facts become clutter, confusing and often meaningless. Without a coherent story, Postman argues, we are overwhelmed by a rush of events, buried by an endless cascade of facts.

We need an overarching story or theory, even if it is mythical, to help us organize facts and make sense of the world. Otherwise we find ourselves perplexed and confused, unable even to recognize what is information and what is not.

In 1962, James Neel, a geneticist at the University of Michigan, offered up such a story about our dietary history. Neel published his theory in an article titled "Diabetes Mellitus: A 'Thrifty' Genotype Rendered Detrimental by "Progress." His conclusions were based on an intriguing assumption, the idea that early humans faced frequent periods of famine, and that those who survived them were endowed with "thrifty genes," which they passed on to their offspring and eventually to modern humans. This supposed adaptation gave many humans a "quick insulin trigger" that increased the ability to store fat from each feast in order to stay alive during the next time of hunger. In today's world of abundance, according to the theory, this adaptation has become a curse, making us more susceptible to obesity, diabetes, and early death.

Neel's theory quickly gained adherents and has been reiterated in countless books and articles over the years. Many authors trying to explain modern obesity in general or why a particular group is obese have invoked it. Phrases such as "feast or famine" and "survival of the fattest" are common in the scientific literature and in popular diet books. Many authors over the years have invoked the theory with an air of surprising certainty, seeming to consider it a proven fact instead of just a theory (Colles 1999, Gaesser 1996).

If Neel is correct, we now are living in a food environment for which our bodies were not designed. Supposedly, our prehistoric ancestors developed thrifty genes to survive frequent severe food shortages. We no longer face such shortages, but our famine-induced thrifty genes are still at work storing fat at a time when it is both unnecessary and detrimental to our health. As a result, millions of modern humans are living in a permanent disease state, with epidemic levels of obesity and diabetes. This might be the price we are paying for living in a modern society — a Faustian bargain at best. But is this theory, which more or less condemns us, actually true?

The thrifty genes theory is perhaps unique in being accepted so widely with so little discussion or debate, but that is not so surprising. The theory had much in its favor. First of all, James Neel was a respected member of the scientific community at the time he proposed it. He went on to have a very distinguished, if sometimes controversial career in genetics. Second, the theory was consistent with what scientists at the time mistakenly assumed about human prehistory. In the early 1960s it was generally believed that Stone Age humans lived a precarious existence in which frequent famines were an inescapable part of life. The idea that human metabolism had evolved to cope with these imagined cycles of feast and famine made perfect sense. Third, as time went on, several isolated populations of humans and animals were found that seemed to exemplify the theory. Fourth, thrifty genes seemed like a very logical way to explain the inexorable rise of obesity that was taking place throughout much of the modern world.

Some have suggested that famine-induced thrifty genes evolved in humans, not while we were living as hunter-gatherers, but after we adopted agriculture (Prentice et al. 2005). John Speakman offers what is perhaps the best argument against this idea and against the thrifty genes theory in general. According to Speakman, if famine had been shaping human metabolism since hunter-gatherer days and thriftiness had a survival advantage (k) of 0.001, then nearly everyone in the world would have thrifty genes by now. As a result, with a few exceptions, the entire populations of industrial societies such as America, Europe, and Japan would be obese. On the other hand, Speakman continues, if famines began to stress human populations only after we adopted peasant agriculture, and assuming once again that metabolic thriftiness had a survival advantage (k) of 0.001, there would not have been sufficient time for thrifty genes to spread widely throughout the world's population. As a result, obesity still would be somewhat rare even in industrial societies.

Wasteful Genes

A paper was published in 2014 titled "On the Evolutionary Origins of Obesity: A New Hypothesis." The authors of this paper challenged Neel's thrifty gene theory and put forth an alternate explanation. They proposed that the differences in metabolic thriftiness we observe among groups or populations are not dietary in origin, but climatic — evolving as the result of early humans leaving tropical Africa and migrating into colder regions such as Europe and Northeast Asia (Sellayah et al. 2014).

In order to survive the prolonged cold they encountered, successive generations evolved the ability to spontaneously burn off calories and generate body heat. This trait, *adaptive thermogenesis*, had significant survival value, and has been passed down to those modern humans whose ancestors came from cold regions. Studies have shown consistently that people whose ancestors came from colder parts of the world, such as Europe and Northeast Asia, have higher metabolic rates than people whose ancestors came from the tropics or subtropics. For example, Americans of European, Chinese, and Korean extraction expend more energy for the same level of physical activity than do Americans who trace their ancestry to Polynesia, sub-Saharan Africa, or South Asia. As a result, Americans of European and Northeast Asian ancestry burn more calories just to get through each day (Gannon et al. 2000).

One study found that European-American women use about seven percent more energy when resting than African-American women. According to the authors, this difference "may contribute to the higher rates of obesity found in the African-American population" (Jakicic and Wing 1998). Adolescent girls showed the same relationship, the resting energy expenditure of Caucasian girls being about 90 kilocalories per day greater than that of African-American girls (Yanovski et al. 1997). These differences did not diminish with age. Caucasian-Americans older than 55 consistently use about ten percent more energy for the same activity than African-Americans of the same age. According to the authors, "Low rates of energy expenditure may be a predisposing factor for obesity, particularly in African-American women" (Carpenter et al. 1998).

Other studies found that European Americans burn more energy both when resting and when active than South Asian or Polynesian subjects. For example, the resting metabolic rate of young Caucasian women was found to be about seven percent higher than that of young Polynesian women (Rush et al. 1997). Other researchers compared the relative energy costs of lying down, sitting, and standing for groups of European, African, and South Asian males. They found that South Asian and African subjects expended about ten percent less energy during each of these activities than Europeans (Geissler and Aldouri 1985).

The research has been very consistent. People of European and Northeast Asian ancestry tend to burn five to ten percent more kilocalories at the same level of activity than ethnic groups whose ancestors came from warmer parts of the globe. In nearly all such comparisons, Europeans and Northeast Asians stand out as being notably unthrifty, and obesity rates in America reflect these metabolic differences. Americans of Polynesian descent have the highest rates of obesity, followed by African-Americans, then Hispanics, and then Europeans, with Northeast Asians coming in as the least obese. According to a 2009 report issued by the Centers for Disease Control, 36 percent of black adults were obese, compared with 29 percent of Hispanics, and 24 percent of whites (Pan et al. 2009). A study published in 2004 showed that ethnic differences were even greater when it came to children. Among adolescent girls, African-Americans had the highest rate of obesity at 26.2 percent, followed by Hispanic girls at 19.4 percent, and then European-American girls at 12.4 percent (Cossrow and Falkner 2004).

Scientists not only have shown that adaptive thermogenesis takes place, they have identified where it occurs in the body; it is in a special kind of tissue called *brown adipose tissue* (BAT) or brown fat. Brown fat is most pronounced in babies and young children, but persists into adulthood, especially in northern Europeans and Northeast Asians.

This puts a whole new spin on the idea of thrifty genes. So-called "thrifty" groups, those whose ancestors came from tropical or subtropical climates, actually represent the "normal" condition in the world. They make up more than 80 percent of the global population, and will account for an even larger share by the end of this century. This suggests that isolated groups with very high rates of obesity and diabetes, such as the Pima Indians and Polynesians, were not really aberrations at all; they simply seemed so in the context of the times. Billions of people living in such places as Africa, Southeast Asia, India, Mexico, and Central and South America are similar to them on the "thriftiness" continuum; and not surprisingly, more of them are becoming obese each year as they migrate to the world's cities and start eating calorie-rich industrial foods.

There is little doubt that inheriting bodies that "waste" calories has given many people of European and Northeast Asian descent an advantage in today's world, making them slightly less prone to becoming obese relative to people of other races. Actually, the advantage may not be so slight. The average adult takes in more than one million kilocalories during the course of year, so a metabolism that is five percent less efficient would result in 50,000 kilocalories being "wasted" yearly, equivalent to about 15 pounds of body fat. But wasting that much energy would have been a dangerous survival strategy in the past. Conditions that drove evolution in such a direction must have been harsh indeed.

The Europeans

Until recently, it was believed that most Europeans were descended from wheat farmers who migrated to the continent from the Middle East during the last 10,000 years as growing populations forced them to seek out new land. Advancing west and north, they supposedly overwhelmed and replaced the scattered bands of Ice Age hunters that had somehow managed to survive the cold for thousands of years. But scientists have recently begun to question this idea after analyzing the mitochondrial

DNA of more than 6,000 modern Europeans. They now believe that most people of European descent (80 percent or more) can trace their ancestry back to the original Ice Age hunters who began migrating to Europe about 40,000 years ago (Sykes 2001).

If these early humans had known what was in store for them, or at least for their descendants, they probably would have gone elsewhere. Perhaps beguiled by an unusually warm period, they picked a very cold and hostile environment in which to settle. As tropical animals, their first mistake was colonizing an area located so far from the equator; for example, much of France and even part of "sunny" Italy are farther north than parts of Canada. In addition to picking a bad place, these early colonizers of Europe picked a bad time as well. They moved onto the European continent just in time to experience the last brutal surges of a long ice age.

Had you visited Europe a little over 100,000 years ago, you would have felt quite at home, since the climate then was much like today, perhaps even warmer. But then a rapid cool down began and by 70,000 years ago Europe had become bitterly cold, with thick ice sheets covering much of the continent. By 60,000 years ago, the continent had warmed somewhat, though conditions still were colder than today. But even this respite soon ended; about 30,000 years ago, the cold once again moved south, bringing ice, snow, and frigid temperatures to most of Europe. This full-fledged ice age lasted until about 14,000 years ago, when the world began to warm up rapidly. Beginning about 13,000 years ago, there was a rapid resurgence of arctic conditions lasting about 1,000 years, but then warm conditions resumed and have continued to this day.

Ice-age Europeans were confronted with a daunting problem. How do tropical animals with no body fur manage to survive and rear children in a climate of long, icy, subzero winters and summers during which the temperature rarely got much above 50 degrees Fahrenheit? Glaciers would repeatedly advance far into Europe during the ensuing centuries, each time forcing desperate bands of humans to seek refuge hundreds of miles to the south. They first had to make it over the Pyrenees into Spain or over the Alps into Italy or perhaps east to Greece before mountain passes were blocked by snow and ice. But they could only go so far before the Mediterranean Sea blocked the way, forcing them to hunker down and endure the cold, usually for many generations.

The mastery of fire and the ability to make fur clothing helped them to survive, but something else was required; their bodies had to adapt to the cold, and eventually they did. Over time, the bodies of warm-blooded animal species can adapt to cold in three ways. One way is to grow longer, thicker fur, which was not an option for early European hunters, since our species had forsaken that strategy a long time ago. Another option is to accumulate thick layers of body fat, but there is a problem with this also. Carrying around a lot of fat slows animals down, making them less efficient at acquiring food and more vulnerable to predators.

The third strategy, one adopted by many cold-adapted animals, is to "waste" or burn off calories for the purpose of generating extra body heat, a process commonly called *adaptive thermogenesis*. Following is a description of this process in experimental animals (Smith and Fairhurst 1958).

> [T]he cold-adapted animal would appear to enjoy a caloric advantage... for the same amount of oxygen consumed and substrate oxidized, the resting heat production of the liver would, on this basis, be doubled in the acclimatized animal...Moreover, if these conclusions can be applied to the whole animal, it may be inferred that the caloric cost of a given work load will be higher in the cold-adapted animal than in the normal.

Scientists have found that animals forced to live in colder areas, such as rodents farther up mountains, burn more energy both when at rest and when active than do members of the same species living in warmer environments. This, according to some scientists, explains the metabolic differences found between Americans of European and northeast Asian descent and those whose ancestors came from Polynesia, Africa, or South Asia. Modern Europeans and Northeast-Asians are much like other animals forced to become cold adapted; under these conditions, their metabolic engines evolved to run hotter than members of the same species whose ancestors remained in tropical climates.

Burning extra calories to generate extra body heat was a good adaptation for surviving in a glacial environment. It now seems to be an advantage to their descendants living in industrial societies awash in sugar, fat, and calories. If your ancestors came from Europe or Northeast Asia, you might have an advantage in an obesity-prone world. In many cases, your body simply burns off an extra 100 or so kilocalories every day. This can be useful for avoiding obesity in a world filled with high-fat, sugar-laden, energy dense foods.

"Then why isn't it working?" one could ask. "The obesity rates of European-Americans might be lower than those of Polynesian, African, or South Asian ancestry, but a lot of European-Americans — about one in three, are obese; and their numbers keep growing just like every other ethnic group. What is going on?"

The answer is quite simple. People of European and Northeast Asian descent do tend to have "unthrifty" metabolisms compared with members of certain other races, but this is little defense against the kinds of high-fat, sugar-rich foods many of them now eat. Americans whose ancestors came from northern Europe or northern Asia might succumb to obesity and diabetes a little more slowly than their fellow citizens whose ancestors came from warmer climates, but in time, all of us will end up in the same condition. The American food environment is so extreme as to overwhelm any inherited metabolic differences. In essence, everyone is in the same boat. Some ethnic groups might be a little more resistant to the modern pandemic of obesity and diabetes than others, but in the long run that means little.

The next chapter examines three populations — the Pima Indians of Arizona, the Polynesians of the South Pacific, and the North African sand rat. Each has been cited as an example of Neel's thrifty gene theory in action — to explain why certain isolated groups supposedly exposed to past famines might have evolved thrifty genes, making them more prone to obesity and diabetes in the modern world. Although they do not really support Neel's theory, and the theory itself is becoming increasingly suspect, examining these three populations can serve a very useful purpose. They can serve as models for all humans, helping us to understand why pandemic levels of obesity and diabetes, once believed to be confined to small isolated groups, now threaten the entire world.

Chapter 2. Pima, Polynesians, and Sand Rats

Three populations have been widely cited as real world examples of how Neel's thrifty genes theory works. These are the Pima Indians of Arizona, Polynesians of the South Pacific, and the North African sand rat. These three groups, although different in many respects, have one thing in common. Shortly after adopting a modern western diet — or a modern laboratory diet in the case of the sand rat, they began to experience extraordinary rates of obesity and diabetes.

Many researchers have attributed the remarkable tendency of these groups to become obese and diabetic to the "fact" that their ancestors experienced frequent episodes of famine. As a result, so the story goes, modern members of these groups have been endowed with very efficient or thrifty metabolisms. When provided with an abundant, dependable food supply, they are compelled to overeat and store fat to survive the next famine. The resulting obesity is inevitable — preordained and programmed by conditions in their ancestral pasts.

There is a major problem with this feast and famine story — lack of objective evidence. The assertions that the Pima, Polynesians, and sand rats all experienced frequent famines in the past are simply assertions, nothing else. No one has ever presented any scientific data to show that they are true. For example, no one to my knowledge has gone to North Africa and observed sand rats in the wild to see if they indeed live a feast or famine existence. The same is true of the Pima Indians. It simply has been assumed (and repeated by numerous authors) that the Pima tribe, just like other "primitive" humans, lived with the constant threat of famine. The assumption seemed so self-evident that almost no one thought it necessary to question it. Surely these poor primitives could not have procured a dependable food supply without the benefits of modern civilization!

Similar assumptions have been made concerning modern Polynesians. Their ancestors supposedly suffered extended famines during the long ocean voyages necessary to populate the scattered islands of the vast Pacific. Modern Polynesians grow fat on a modern diet because they are descended from the survivors — those whose thrifty metabolisms enabled them to stay alive on these long journeys. However, as in the case of the Pima and the sand rats, these assertions are pure conjecture. No data or historical records have been offered to back them up.

The Sand Rat

The sand rat lives in the deserts of North Africa and Arabia. It makes its home at the base of a saltbush, from which it gets all of its requirements for life. The leaves of this plant provide all of the sand rat's food as well as its water — but not without considerable work. The sap of the saltbush is saltier than seawater and the leaves are very low in nutritional content. The sand rat spends most of its time eating. It chews off a leaf, removes the tough, salty outer layer with its teeth, and then chews and swallows the fleshy inner part. The busy little rodent repeats this procedure again and again. By the end of the day it has eaten as much as half or more of its body weight in saltbush leaves. A human adult of average size would have to consume more than 20 pounds of food at each meal — breakfast, lunch and dinner — a total of more than 60 pounds a day, to equal this amazing animal in pure eating power.

More than 50 years ago, Duke University scientists imported some sand rats to study in their laboratory. They wanted to find out how the little creatures were able to tolerate such high levels of salt in their diet. Upon feeding them normal rat chow, the researchers were amazed by the transformation that took place. The animals became so fat that their bellies dragged the ground. Many developed diabetes early in life and died at an early age. Most who managed to live long enough were blinded by cataracts. Eating rat chow provided calories in a much more concentrated form that the sand rats were used to; when they ate anything near the volume of food they normally consume, their systems went out of control.

The saltbush provides the sand rat with a dependable, though extremely bulky and low-energy diet. Over generations these resourceful rodents have learned to thrive on the equivalent of tough lettuce saturated with salt water. They have to eat an incredible volume of saltbush leaves each day to meet their energy and nutrition needs. When introduced to commercial rat chow, sand rats have no way of sensing that their new rations are many times more energy dense than the food their bodies were attuned to. They continue to eat large amounts of food. This soon makes them diabetic and obese (Bennett and Gurin 1982, Pool 2001).

One does not have to invoke past famine to explain this. The North African sand rat has evolved an extremely thrifty metabolism not in response to famine, but in order to cope with a food supply that is extremely bulky and low in calories. Notably, the Duke scientists did not solve their problem by instituting a feeding schedule that included periodic famine. Instead, they simply fed the sand rats a low-energy diet consisting exclusively of fresh vegetables such as lettuce and celery.

The Pima

The Pima, or the *Akimel O'otham* (River People), as they once called themselves, have lived for centuries along the Gila River in Arizona. They used water from the river to irrigate fields of corn, squash and beans. They harvested wild plant foods such as honey mesquite pods, cholla-cactus buds, poverty weed and prickly pears from the surrounding desert. They caught fish in the Gila River and hunted game animals such as mule deer, jackrabbits and doves. Their environment provided them with an austere but dependable food supply. The Spanish conquistadors visited the Pima lands around 1530. Their accounts indicate that the tribe was prosperous and friendly, providing their visitors with both food and guides (Howard 1997).

The Pima seemed equally prosperous when American pioneers passed through their lands 300 years later on their way to California. The friendly tribe gave the travelers food and other assistance. Legendary army scout Kit Carson, who visited the Pima in 1846, reported that tribal members welcomed him and generously shared their food (Pool 2001). A United States Army battalion also visited the tribe in 1846 (Griffin 1943). Their surgeon, Dr. John Griffin, described the Pima as being "sprightly" and in "fine health." He also observed that the tribe had "the greatest abundance of food, and take care of it well, as we saw many of their storehouses full of pumpkins, melons, corn..."

Anthropologist Frank Russell lived with the Pima for several years around 1900 and wrote the definitive text on the tribe. He reported that in the past, nearby tribes who were less fortunate, notably the Apaches, attacked the Pima regularly, sending out a small war party every few days and larger forces every month or so. The Pima responded to these constant attacks by becoming efficient, formidable warriors. These desert dwellers were fit and tough. Mercifully, they could not foresee the obesity, diabetes and disability that would begin to overwhelm them in the 20th century. Things started to go bad for the tribe around 1860, when settlers established farms upstream and began to take irrigation water from the river. Soon additional upstream water was diverted for use in mines and smelters.

By 1900, the economy and culture of the Pima had changed forever. They continued to farm, but the supply of irrigation water had become unpredictable, and they were increasingly dependent on food donated by the U.S. government or purchased from local trading posts. Refined flour, sugar, lard and canned meats became dietary staples and these new foods began to remake the Pima people. According to Russell (1975), a visitor in 1902 reported that tribal members "exhibit a degree of obesity that is in striking contrast to the 'tall and sinewy' Indian conventionalized in popular thought." The traditional Pima diet contained about 15 percent fat. By 1950 this had increased to 24 percent and by 1970 to nearly 45 percent. In addition to more fat, the Pima ate more sugar and more refined flour. Their fiber intake declined dramatically. The consequences soon became apparent to the outside world.

In 1963 a team from the National Institutes of Health arrived to study the Pima. They wanted to compare the prevalence of rheumatoid arthritis among tribal members with that among the Blackfoot tribe in Montana. However, they found something else — one of the highest rates of obesity and diabetes in the world. Two years later they returned to focus on this striking phenomenon. Government scientists began examining thousands of Pima each year, measuring height, weight, blood pressure, and blood sugar. They also assessed kidney function and examined eyes to check for vascular damage from diabetes. And they wrote research papers — lots of them. Publications resulting from studying the tribe reportedly take up 30 to 40 feet of shelf space at the National Institutes of Health library in Phoenix (Gladwell 1998).

These studies unearthed a tragic situation. The average Pima man in his early 30s was about five feet, seven inches tall and weighed 220 pounds. Women were even heavier. The average Pima woman was slightly over five feet, two inches tall and weighed 200 pounds. Some tribal members weighed as much as 500 pounds. Diabetes had become an epidemic. In 1908 a visiting doctor/anthropologist had recorded only one case of diabetes on the reservation. In 1937 there were only 21. In 1954, 283 cases were identified, a ten-fold increase. The Pima now have one of the highest rates of diabetes in the world. Half of all adults over the age of 35 are afflicted; this is five times the rate in the rest of the United States. Things can only get worse for the next generation. Pima children now are becoming obese and diabetic at earlier ages. Rates of obesity and diabetes among future adults likely will surge even higher (Bennett 1999).

The thrifty genotype theory is invoked to explain the predicaments of both the Pima and the sand rats. It is assumed that famine was a constant threat to both groups. It is further assumed that individual sand rats and Pima who had the most efficient metabolisms were most likely to have survived the frequent periods of starvation they supposedly had to face. These lucky few would have stored up fat during times of plenty to get them through the next famine, during which their less fortunate brethren would be dying all around them.

Those unable to withstand the famines did not leave descendants. They or their children died off in the periodic famines because their metabolisms were not thrifty enough to store enough extra fat during times of plenty. They were replaced by those whose bodies were better able to stash away extra calories in the good times. Now, year after year, modern Pima are getting fatter, hoarding extra calories in preparation for the famine that will never come.

But there is a major flaw in this interpretation. To my knowledge, no archeologist, historian, or other specialist has documented any of these "periodic famines" in the Pima past. The arid landscape they inhabit might look forbidding to visitors from more humid, well-watered regions, but their desert home provided them with a dependable food supply for generations. They harvested a variety of different foods from four independent sources: fish from the Gila River, cultivated crops from irrigated fields, and both wild plants and wild game from the surrounding country. This redundancy insured that the Pima almost always would have something to eat.

For famine to ensue, they would have to be very unlucky. Several unfavorable events would have to occur in the same year. First, the irrigated crops would have to fail. Then, upon visiting the surrounding desert and mountains, the Pima would have to find that the native plants had produced little or no edible fruits, seeds, or roots. Even more bad luck would have to follow. Upland animals would have to vacate the surrounding lands and fish would have to disappear from the river. Rarely would all of these food sources fail at the same time. A mixed economy of farming supplemented by hunting and gathering typically produces a high level of food security. It is likely that throughout their history the Pima easily procured more food than they needed in nearly all years.

However, the traditional diet of the Pima, although generally reliable and nutritious, was very low in energy density. They faced the same biological challenge as the sand rat — adapting to a diet that was very low in calories per given volume. The Pima, relatively isolated in their desert valley, came to thrive on the Spartan menu that nature had provided them. In each generation, those individuals who were more efficient at processing and extracting calories from a large volume of food low in energy density were more likely to produce healthy offspring.

The advantage did not have to be dramatic. Those with thriftier metabolisms had only to leave a few more offspring in each generation. The process would have a powerful cumulative effect — like compound interest. Eventually, their progeny would come to dominate the population. Both the Pima and the sand rats adapted to challenging environments by developing thrifty metabolisms. But this "thriftiness" was not a response to famine. Instead, each group developed this trait because their ancestors were forced to live on a generally reliable, but very bulky, energy-poor diet for hundreds of generations.

There is another interesting chapter to the Pima story. A little less than 1,000 years ago, the tribe split apart. One group stayed in Arizona; the other band made its way into the remote mountains of northern Mexico. Genetically, the two groups are similar, but the Mexican Pima, unlike those who remained in Arizona, had not yet adopted modern food or a modern lifestyle. The Mexican Pima continued to grow corn, beans, potatoes, and other vegetables on small plots, cultivating by hand. They also kept some livestock.

The Mexican Pima remained much like their Arizona relatives had been before they started eating the modern American diet. They were shorter than modern Arizona Pima by about two inches and weighed much less. A sample of middle-aged Pima men from Mexico weighed on average about 150 pounds. A similar group of Pima men living in Arizona weighed nearly 200 pounds. Similar comparisons were done for women. Pima women in Arizona weighed about 50 pounds more than their Mexican counterparts.

How does one explain these striking differences? There is no documented history of famine among the healthier, thinner Mexican Pima. Why were they not obese? Should they not have been getting fat in anticipation of the next period of starvation? The answer is that most of them could not. The energy density of the food they ate and the moderate level of physical activity that was integral to their lives made becoming fat an almost impossible task. The Mexican Pima did not have ready access to such American delicacies as greasy fast food, candy bars, cookies, or high-fat, high-sugar glazed donuts. They do not drink sugar-laden, carbonated soft drinks every day. Instead, they ate mostly natural foods that contain little sugar or other highly refined carbohydrates. Their food was low in fat content and overall energy density while being high in fiber.

It is likely that this diet was similar in calorie and fiber content to that eaten by their ancestors prior to the arrival of Europeans. The Mexican Pima walked every day and performed at least moderate amounts of manual labor. This is why they stayed lean and healthy while their Arizona cousins became fat and sick. It would be interesting to know if the Mexican Pima have retained their traditional way of life and remained healthy or fallen prey to the lures and afflictions of modernity.

The Polynesians

Polynesians are another group believed to exemplify the thrifty gene theory in action. Few Polynesians were overweight or obese when Europeans first encountered them. Regardless of the island they visited, early seafarers were struck by the stature and physical attractiveness of the natives living there. Many travelers made note of this in their journals, as illustrated by the examples below.

> "The inhabitants of these islands, collectively, are without exception the finest race of people in this sea. For good shapes and regular features, they perhaps surpass all other nations." Captain James Cook, English Explorer (Cook 1784)

> "They are above the common stature of the human race, seldom less than five feet eleven inches, but most commonly six feet two or three inches, and in every way proportioned." Commodore David Porter, United States Navy (Porter 1823)

> "In form of beauty they surpassed anything I had ever seen...nearly every individual of their midst might have been taken for a sculptor's model." Herman Melville, American author (Melville 1846)

But these handsome, healthy people were doomed. The seafaring adventurers from America and Europe brought death, disease, and degradation with them. The Pacific Island culture soon was destroyed and the natives were devastated by western diseases such as smallpox, influenza, and tuberculosis. Only a few survived to leave descendants. These descendants now face a new curse — pandemic levels of obesity and diabetes. Nauru, a tiny spit of land only eight square miles, is a striking example.

For thousands of years seabirds congregated on the island, literally burying much of it under deep deposits of phosphate-rich guano.

The 8,000 or so natives became wealthy by selling the deposits to fertilizer companies. Their new-found wealth enabled them to abandon their traditional diet of fish and vegetables. They exchanged it for one of store-bought foods high in fat, sugar and refined flour. They no longer needed to work, so many sat around for much of the day eating calorie-rich food and drinking beer. The people of Nauru now have the distinction of being the fattest in the world, with an obesity rate of 61 percent.

This modern curse of obesity and diabetes has consumed island after island in the south Pacific. A western dentist who lived on the Micronesian island of Kosrae painted a grim picture of both adults and children. He observed that children were given candy to keep them quiet in church and that younger children were nursed with sugar water. Calcium and vitamin A deficiencies were common and probably half the teenage girls were obese. The native diet consisted largely of imported foods, high in refined flour, fat, and sugar. Many of the adults suffered from diabetes or other chronic illnesses and leg amputations were common. But, the dentist wrote at the time, "it's what they're used to." In addition to Kosrae, the native people of Hawaii, Samoa, New Zealand, Nauru — nearly all of the Pacific islands, are suffering the same ills. Descendants of a race once renowned for their beauty now are used as exemplars of runaway obesity (Shell 2002).

It commonly is argued that Polynesians, like the Pima Indians and sand rats, are obese and diabetic on a modern diet because they too have inherited thrifty genes. Numerous authors have proclaimed that the ancestors of modern Polynesians experienced severe episodes of famine as they made their long, arduous journeys across the Pacific in search of islands to colonize. They theorize that individuals most likely to survive were those who began with a large store of fat or were better able to withstand starvation because of an efficient metabolism. A severe episode of genetic selection supposedly occurred during this colonization period, favoring those with thrifty genes. As described by Bennett and Gurin (1982) the ancient Polynesians passed "through the eye of an evolutionary needle." If true, this scenario places modern Polynesians in the same predicament as the Arizona Pima. A thrifty metabolism that enabled their ancestors to survive periodic famine has become a curse in the modern world.

But, as in the case of the Pima, the famines that supposedly endowed today's Pacific islanders with thrifty genes are purely conjecture. There is no archeological evidence to that effect and the oral histories of the islanders do not recount times of great hunger in the past. Early ships' journals did not describe the natives of any island as being hungry or starving. They invariably described them as healthy and well-nourished. Despite having plenty to eat, the natives were almost universally lean; none was described as fat. Although food was abundant, no one seemed to be storing excess fat in preparation for the next famine.

Although past famine is not the culprit, the obesity suffered by modern Polynesians most likely was nurtured in their ancestral past. Native Pacific Islanders are descended from the Lapita people. Several thousand years ago, these skilled seafarers began to colonize the far-flung islands of the Pacific. Before beginning their explorations, they had learned the fine arts of boat construction and seamanship as ocean-going obsidian traders. Obsidian is a glassy volcanic rock that was highly prized during the Stone Age, since it could be worked to a very sharp cutting edge.

The islands to which the ancient Polynesians sailed were biologically impoverished. Few plants from the continents could spread to isolated bits of land distributed over such a vast area. As a result, the islands had almost no native plant species that could be used for food. The colonizers had to bring their agriculture with them. They stocked their canoes with plants such as taro, breadfruit, yams, bananas, coconuts and bamboo. Some brought pigs, dogs or chickens. Many islands ended up with only one domestic mammal — for example, only dogs or only chickens. Most islands had an abundance of native turtles, fish and shellfish, as well as large flightless birds that had never encountered a predator. Most of the birds soon were harvested to extinction. Archaeological digs show that initially the colonizers ate large numbers of the native animals, including birds, turtles, shellfish and fish. But as time went by these native food sources became scarce and islanders relied increasingly on the domesticated animals they had brought with them. The fruits and vegetable plants they had imported also were important food sources (Dye and Steadman 1990).

The early Polynesians had an abundant and diverse diet, but one that was low in energy density. The birds, turtles, shellfish and fish that they ate were low in fat and the vegetables, and fruits they cultivated were low in calories and high in fiber. Like the Pima, most pre-contact Polynesians could not have become fat even if they wanted to. The bulk and low energy density of their food supply would not allow it. The diet of early Hawaiians, for example, has been well documented. More than 60 percent of their calories were derived from the taro root and sweet potato. These staples were supplemented with coconut, breadfruit, bananas, papaya, guava, fish and pig. This traditional diet typically contained less than ten percent fat and a lot of fiber.

Many native Hawaiians have become very obese on modern foods. A Hawaiian physician, Dr. Terry Shintani, and his collaborators have demonstrated that obese native Hawaiians can lose impressive amounts of weight by consuming a diet similar to that of their ancestors. University of Hawaii researchers persuaded a group of obese individuals to try a traditional Hawaiian diet for 20 days. They were allowed to eat all of the taro, breadfruit, yams, sweet potatoes, taro leaves, seaweed, and fruits they wanted. Researchers put limits on the amounts of chicken and fish the participants could eat. All food was eaten either raw or steamed, as was done traditionally. Participants consumed as much as four or more pounds of food a day and still lost an average of 17 pounds during the study. Their energy intake declined from more than 3,000 kilocalories a day to fewer than 1,600 kilocalories (Shintani et al. 1991).

Both the Polynesians and the Pima Indians (and the little sand rat) do have thrifty metabolisms which makes them unusually susceptible to obesity. However, there is little evidence that past famine is the cause. Instead, each of these populations inherited thrifty metabolisms because their ancestors lived for many generations on a diet that was dependable but very high in bulk and low in energy density.

At the time the Pima and Polynesians were being studied, obesity rates still were relatively low in America. In the 1950s many children still walked or rode their bikes to school, and although some people owned televisions, there were few channels and the variety of programming was limited, so kids did not spend hours glued to the screen. Most people did not yet have air conditioning and neighborhoods were safer, so children spent more time playing outside, especially during the summer. Many Americans were still performing manual labor in factories, at construction sites, and on farms and ranches. A lot of construction work was still done with shovels; most ships, trucks and train cars were still loaded and unloaded by hand; and in 1950, field laborers were still picking 90 percent of America's cotton crop. Fast food restaurants were few and far between, and machines dispensing snack foods and sugary soft drinks were not yet in every commercial and public building, including schools. Soft drinks were still considered a treat rather than something to be consumed daily as a replacement for milk or water.

America's obesity rates reflected this 1950s way of life. Only about ten percent of Americans were obese then, and morbid obesity, now a common theme of television documentaries and reality shows, was extremely rare. Overweight and obesity were even less common in Europe, which had not fully recovered from the ravages of World War II, and were almost nonexistent in such places as China, Japan, Indonesia, Africa, Mexico and Central and South America. In a world where obesity remained something of an oddity, it was easy to assume that groups such as Pima and Polynesians must be metabolic outliers, somehow different from the rest of humanity. It also seemed logical that the "thrifty genes" responsible for their condition would prove to be relatively rare. But they were not rare; instead, so-called thrifty genes turned out to be the norm.

Ancestral Humans

Modern humans first appeared on Earth between 150,000 and 200,000 years ago. By "modern humans" I mean people who looked like us and had similar levels of intelligence. They could pass the subway test. Assume you somehow transported a man from 100,000 years ago to the present, dressed him in modern clothes and had him get on a crowded subway car. The other passengers would not give him a second glance. Early humans also were similar to us in intelligence. Assume one could somehow transport a group of infants from 100,000 years ago to the present. Assume further that modern couples adopted them and eventually enrolled them in school. These children from the past soon would learn to read, write, and do math. The point is, people walking the earth a thousand or more centuries ago were much like us; only their way of life was different.

Our human ancestors adopted agriculture just in the last 10,000 years. For all the time prior to that, more than 95 percent of their tenure on earth, they lived as hunter-gatherers. There are no written accounts of this vast expanse of time — hence the term "prehistoric." Anthropologists have gained some knowledge of early human life by uncovering the remains of old settlements and campsites. Discarded tools, weapons, animal bones, seeds, ornaments — all have provided insight into how early humans lived. Additional information has come from the study of the few isolated groups of hunter-gatherers that somehow survived into modern times.

Long before any studies were made of hunter-gatherer life, the conventional wisdom held that our existence prior to agriculture was one of misery and starvation. This somber view first entered the human consciousness more than 300 years ago. Thomas Hobbes gave it public voice in 1651, when he penned his famous description of pre-agricultural humans, "No arts, no letters, no society, and, which is worst of all, continual fear and danger of violent death and the life of man solitary, poor, nasty, brutish, and short." This description portrays early man as a hapless, ineffectual creature living constantly on the edge of starvation.

Hobbes had an agenda; his aim was to extol the value of civil rule and political control in improving the quality of human life. But he was living in the Middle Ages, a time when the populace was ravaged by famine, disease, and political violence. If Europeans in the 1600s were to be convinced that they were blessed, then life before "civilization" and rulers had to be portrayed as a nightmare.

The inherent "logic" of Hobbes' assertion seemed unassailable to most of his contemporaries. He was writing at a time when nearly all Europeans were devout Christians who rarely questioned religious doctrine. They were convinced that God had a plan for the world and that this plan included the inevitable improvement of the human condition. Considering this, life in Europe during the 1600s, despite its horrors, still had to be a vast improvement over the Hobbesian hell in which uncivilized humans once had been forced to live. Somewhat surprisingly, this religion based reasoning later drew support from Darwin's theory of evolution, which was widely misinterpreted to make the same case. Herbert Spencer, a young contemporary of Charles Darwin, is credited with coining the phrase "survival of the fittest." He believed that evolution reflected a plan pre-ordained by God and thus was moving in a purposeful direction.

The idea that humans lived a precarious, hungry existence for most of our history was an accepted doctrine among mainstream anthropologists until recently. A once widely-used text, Braidwood's *Prehistoric Men*, first published in 1948, categorized hunter-gatherer life simply as "a savage's existence." This same text offered the opinion that a human who spent his whole life following animals and trying to kill them or wandering haplessly in search of plant food was not really a human — but was more like an animal. Unfortunately, many of the authorities making such declarations and writing books at the time had never seen a hunter-gatherer in the flesh.

The prevailing negative view of hunter-gatherer life began to change after 1960 when researchers actually started living with some of the surviving groups and observing their day-to-day activities. Immediately prior to the adoption of agriculture, there were an estimated five to ten million hunter-gatherers in the world, occupying nearly every part of the land not covered with desert or permanent ice. By 1960 their numbers had shrunk to around 30,000. Small self-sufficient groups were still living directly off of the land at remote locations in Australia, Africa, South America, Thailand, India, Greenland, and the Northwest Territory of Canada (Pfeiffer 1969). Fortunately, anthropologists were able to study some of these surviving bands before their old ways were gone forever.

The aborigines of Australia and the Bushmen of southern Africa have been studied most extensively. In both cases, real studies of actual people showed that most of the conventional wisdom held for so long was wrong. Close observation revealed that these "primitive" groups were well-supplied with food year-round and were able to procure it with surprisingly little work.

This was not a revelation to some people. English explorers traveling in the Australian outback as early as the 1800s had reported that the aborigines enjoyed a leisurely life with an abundant, reliable food supply. Sir George Grey (1841) explored remote areas of Australia during the years 1837 to 1839. He described aborigine subsistence, "In all ordinary seasons, they can obtain, in 2 or 3 hours, a sufficient supply of food for the day, but their usual custom is to roam indolently from spot to spot, lazily collecting it as they wander along." Another British explorer (Eyre 1845) traversed much of central Australia in the years 1840 to 1841. He wrote, "I have found that the natives could usually, in three or four hours, procure as much food as would last for the day, and that without fatigue or labour."

Such observations were ignored or discounted for many years, since they were contrary to the perceived wisdom. But eventually they were confirmed by modern research. Once such study reported that Australian aborigines spent only four to five hours each day gathering and preparing food, and this time included frequent stops for naps or conversation. The aborigines easily obtained sufficient calories and an abundance of protein with only a moderate amount of work (McCarthy and McArthur 1960).

The Bushmen of southern Africa have a lifestyle very similar to that of the Australian aborigines. They occupy arid and semi-arid parts of Botswana and South Africa that typically get less than nine inches of rain each year. The forbidding terrain they inhabit yields a surprising variety of food resources. In one area that was studied intensively, the Bushmen were able to choose from more than 80 species of plant foods. This included 29 species of fruits, berries and melons and 30 species of roots and bulbs. Their hunting grounds contained more than 220 animal species, but the Bushmen were selective. They considered only 54 of the species as edible and hunted only 17 species on a regular basis.

Although they usually had many foods to choose from, the mongongo nut was the Bushmen's staple food. It is abundant and nutritious, containing five times the calories and ten times the protein of cereals like wheat of corn. This one food made up about 30 percent of the Bushman diet. One Bushman reasoned that it would make no sense for his people to farm because there were so many mongongo nuts in the world.

Bushmen usually located their home or base camp near a permanent water hole. From a central location they could exploit an area up to six miles distant in all directions. This is about as far as they were willing to walk and return the same day. A six-mile radius gave them access to one hundred square miles of territory. They moved their home camp several times during the year to utilize more distant food sources, notably more distant mongongo nut forests, but they rarely traveled very far.

Richard Lee made a thorough field study of Bushman life from 1963 to 1965. He closely monitored the time they spent hunting animals and gathering plant foods. He found that food was available throughout the year and it took only moderate effort and surprisingly little time to procure it. During some seasons, notably spring, the Bushmen had to cover a larger area to meet their energy and nutritional needs. But they never starved or even suffered serious hunger. They managed this with only 65 percent of the population actively involved in hunting or gathering plant foods. The rest were either too young or too old. Even more amazing, this 65 percent worked only about 35 percent of the time. Active workers were able to support themselves and any dependents while working on average fewer than three days per week (Lee 1998).

Although things might get a little lean during some parts of dry years, famine was unknown among the Bushmen. Obesity also was unknown. Although Bushmen rarely experienced real hunger, their food supply and lifestyle would not permit them to get fat either. They remained constantly lean. Harold Thomas, a Harvard physicist, calculated the maximum calories per person the Bushmen could extract from their environment. He determined that they could never exceed 3,200 kilocalories per person no matter how hard they worked. Most of these extra calories would be spent hunting, gathering and processing the extra food, so the extra effort would be wasted. Any weight increase would be minimal. The same would be true of Australian aborigines or any other hunter-gatherer society. Their lifestyle and the nature of their food supply would preclude significant weight gain (Pfeiffer 1969).

Australian aborigines and African Bushmen are instructive examples. Both are small remnants of hunter-gatherer populations that once occupied much larger territories. In both cases powerful, aggressive invaders forced them onto marginal lands that were viewed as not worth taking. Yet these dispossessed peoples adapted to their new habitats and learned to survive.

If the aborigines and Bushmen can wrest a dependable living from such forbidding terrain, think how well their ancestors must have fared when they occupied richer, more productive lands. The aborigines once had the run of all Australia and the Bushmen once roamed throughout much of sub-Saharan Africa. One must assume that such clever, resourceful people coped very well, that famine and want were almost unknown to them.

Another group of hunter-gatherers, the Hadza, live in Tanzania, about 1,000 miles away from the Bushmen. Interestingly, their speech uses many of the same distinctive clicking sounds. They live in a richer environment and their lifestyle is believed to more closely approximate that of most early humans. Studies have shown that they can secure an abundance of food while working only two to three days a week. Just as with the Bushmen, famine is outside their experience. There has always been an abundance of food, so they make no effort to store any for the future, not even for the next day.

They fared very well during the severe drought that hit Tanzania in the 1970s; they were largely unaffected by the crop failures, dying cattle, and widespread hunger that devastated their agricultural neighbors. Some members of nearby farming communities weathered the crisis by joining the Hadza and becoming hunter-gatherers for a while (Crawford and Crawford 1972). Many other examples could be cited to show that nearly all hunter-gatherer societies managed to procure an abundant food supply throughout the year. They were able to achieve this with only moderate effort while enjoying an enviable amount of leisure. By all accounts, hunter-gatherers were some of the best fed and healthiest people on Earth (Farb 1978).

In the fall of 1999, I was collecting soil samples in rural Denmark to be analyzed as part of a world soils database. Along with some Danish soil scientists and geographers, I spent a few hours visiting the Moesgaard Museum in the town of Aarhus. Much of this museum was devoted to the remains and relics of what my Danish colleagues referred to as "ancient Danes." Anthropologists working in the country have located and preserved the bones and artifacts left by many early inhabitants. I was struck by the following inscription on one of the display cases: "Bones of more than 1,000 stone age individuals have been found and examined... Traces of illness caused by dietary deficiency or infection are very rare."

The most important assumption underlying Neel's thrifty genes theory appears to be unfounded. This is the idea that many modern humans have inherited thrifty metabolisms because their ancestors were subject to frequent famines. Convincing evidence of this is lacking. Instead, what we now know about early humans point to a life in which food usually was abundant and obtaining a dependable, healthy diet required only moderate effort. The next chapter looks more deeply into how the human body has evolved to metabolize food and store energy.

Chapter 3. Starvation and Overfeeding

Ancel Keys was a towering figure in the field of nutrition research. He died in 2004 at the age of 100 after a long and illustrious career. He was a living rebuttal to the common view that scientists and scholars are boring and unadventurous. Keys was born in Colorado Springs, Colorado, in 1904 to teenage parents. The family moved to California while he was still very young and eventually settled in Berkeley. From his early years, Keys showed the compulsive, impetuous character often associated with the creative intellect. As a teenager he ran away from home and worked in a gold mine in Colorado, in a lumber camp, and even spent some time shoveling bat guano in Arizona caves. He married for the first time at the age of nineteen, but soon he was divorced.

His early career path took many sharp turns, reflecting his broad interests and quixotic nature. He began the study of chemistry at Berkeley, but soon dropped out to take a job on a ship bound for China. He later returned to Berkeley and earned a BA degree in economics and political science. He then worked for a while as a management trainee at a Woolworth's store. Not surprisingly, Keys found this career not to his liking. He returned to Berkeley and earned a degree in zoology in only six months. He then enrolled in graduate school and earned a doctorate in oceanography and biology at the age of 26. He later studied at Kings College, Cambridge, where he earned another Ph.D., this time in physiology.

Keys eventually took a position at the University of Minnesota, where he established what came to be known as the Laboratory of Physiological Hygiene. He was keenly interested in how the human body reacts to extreme conditions. He once spent ten days at an elevation of 20,000 feet in the Andes to measure the effects of high altitude on his own blood. He repeated this experiment 30 years later with five members from his original expedition. The aim was to see how elderly men reacted to high altitude.

During World War II the government asked Keys to develop compact, light-weight food packs for paratroopers. Keys' efforts resulted in the now famous K rations. He always maintained that the letter K printed on each package was in his honor. He later studied the relationship between dietary fat and heart disease and earned such recognition for this work that he came to be called "Mr. Cholesterol." His influence on dietary science was so notable that he was featured on the cover of the January 13, 1961 issue of Time Magazine.

Keys conducted another war-related study which has been widely cited — a project to determine the physical and psychological effects of human hunger. In the late fall of 1944, he assembled 36 healthy young men on the University of Minnesota campus for a long-term starvation experiment. Their average age was a little over 25. They were mostly members of the Church of the Brethren or the Society of Friends and all of them had refused to take part in the war because of their religious beliefs. Although secure in their convictions, they apparently felt some guilt about avoiding combat while so many of their contemporaries were fighting and dying in the war. As a result, they readily subjected themselves to a prolonged hunger experiment. The possibility that such a study might help starving people in Europe undoubtedly influenced such idealistic young men. Keys hoped his research would provide knowledge that could be used in treating starving and sickly people liberated from war-ravaged areas of Europe and elsewhere.

During the first three months of their stay at the University, the young men's diets were not altered. They continued eating their accustomed amount of food, averaging about 3,500 kilocalories per day. Each man was required to walk about three miles every day and walk on a treadmill for 30 minutes one day each week. At the end of three months, their physical activity was kept the same, but their energy intake was cut roughly in half — to about 1,600 kilocalories per day. The nature of their diet also was changed. Their new meals consisted mostly of whole wheat bread, grains, potatoes, turnips, and cabbage. This was deliberate; when hungry people in war-torn Europe were able to get something to eat, it usually was some combination of these peasant foods.

The young men began losing weight rapidly. Within a couple of months they lost half of their total body fat and some of their muscle mass, taking on the emaciated look of famine victims. They were cold all the time and their skin had a bluish tinge. Small cuts were slow to heal. Some of the men lost significant amounts of hair. They became irritable and lethargic, and ceased to care about their personal appearance. The men avoided as much activity as possible. Two subjects suffered emotional breakdowns and did not complete the study. One of the men chopped off the tip of his finger, perhaps in the hope of getting out of the study.

By the end of the starvation phase, the volunteers had lost one-fourth of their beginning weight. Keys then began the process of refeeding. He restricted their intake at first, but then allowed them to eat as much as they wanted. When set free, the men ate ravenously, but remained constantly hungry. Despite the fact that they now were consuming an average of 5,000 kilocalories per day, they remained obsessed with food. Eating voraciously did not assuage their constant hunger. Their bodies seemed to "remember" how much fat they were "entitled" to, and would not give them any rest until it was restored. They stopped their frantic eating only after they had regained their approximate starting weight. This experiment resulted in a two-volume book, *The Biology of Human Starvation* (Keys et al. 1950), which remains the most comprehensive scientific account of how humans react to prolonged food deprivation.

Fattening Up the Prisoners

Ethan Allen Sims, a Vermont professor, was in many ways the opposite of Ancel Keys. Unlike the flamboyant Keys, Sims has been described as unassuming, "a wiry, soft-spoken New Englander" (Bennett and Gurin 1982). In contrast to Keys' focus on starvation, Sims was interested in how humans of normal weight would respond to deliberate and prolonged overfeeding. He began a series of experiments in the 1960s to address this question, starting out with a small group of student volunteers. The students agreed to deliberately overeat for a designated number of days with the goal of increasing their body weights by 20 percent. Despite consuming enormous amounts of food, they had a difficult time gaining weight. Somehow their bodies seemed to actively resist putting on fat. It soon became clear to Sims that, for normally lean young people, deliberately getting fat by overeating would require great effort. It might well become, in Sims' words, a "full-time job" (Sims 1974).

Then Sims had an inspiration. He found the ideal setting and the ideal subjects to expand his studies. In 1964, the warden of the Vermont State Prison agreed to let a group of prisoner volunteers take part in an overeating experiment. They all were lean young men at the beginning of the study. The warden cooperated in setting up a special research area at the prison containing a kitchen, dining room, and recreation area. The prisoners then began their efforts. For 200 days they ate voluminous amounts of food, essentially doubling their normal intake of calories. Their goal was to gain 20 to 30 pounds. Like the students before them, nearly all of the men found it extremely difficult to put on weight. Most of them eventually managed to gain the targeted amount of 20 to 25 pounds, but they had to eat huge amounts of food to do so.

After reaching their goals with much difficulty, the prisoners found it necessary to consume an average of 2,000 more kilocalories each day than usual just to maintain the added weight. One prisoner gained 28 pounds, but he could do so only by eating 7,000 extra kilocalories a day for the last two months of the study. Unlimited eating might seem an enjoyable way to spend your time in prison, but the novelty wore off very quickly. Most of the prisoners soon found the process unpleasant.

Before long, many had trouble eating breakfast and some would vomit immediately after finishing the meal. In addition to suffering nausea, many of the young men became listless and apathetic. When the experiment ended, they all drastically reduced their food intake and lost weight rapidly. Within a short time, nearly all of them had returned to their original weights, give or take a few pounds. Sims noted the similarity between deliberately overeating and starving oneself, stressing that in both cases the subjects soon returned to their normal weights when allowed to eat normally again. *Guru Walla*

Pasquet et al. (1992) described an unofficial overfeeding experiment carried out by young men in the African country of Cameroon, a ritual called *Guru Walla*. Over a period of several months they ate and ate, trying to put on as much weight as possible. They were driven to do this because the individual who became fattest in the shortest time gained enormous prestige. Their bodies responded to this massive overfeeding by greatly increasing their metabolic rates, by as much as 50 percent. Despite burning off a lot of the excess calories, the men still managed to gain weight, often more than 35 pounds each — but only temporarily. When the ritual was over and the men began eating normally again, they shed the pounds rapidly. Soon they were back at their starting weights, ready for next year's *Guru Walla*.

Taken together, these experiences with overfeeding and underfeeding provide some clear insights into how the human body regulates body weight and eating behavior. First, the thing actually being regulated is not eating; it is the level of body fat. Eating behavior — increasing or decreasing calorie intake, seems to be just a means to that end. Keys showed that people become very distressed when their fat is depleted below the amount their body "thinks" it is entitled to. When the starvation ends, they are compelled to eat voraciously until fat levels are restored. This is no surprise to legions of dieters, who find weight-loss efforts to be futile because the pounds they shed with so much dedication and effort almost always come back. But the response of the Vermont prisoners to massive overfeeding was a surprise to many. Having difficulty losing weight is to be expected, but difficulty gaining weight is another matter entirely. Haven't our thrifty genes programmed humans to compulsively overeat and gain weight anytime we have access to extra food?

A very clever researcher named Rudy Leibel and his colleagues conducted some classic studies which explained much of what was going on. They looked at both sides of the coin — how human metabolism responds to either undereating or overeating for extended periods of time. They were able to sequester student volunteers in a university hospital for months at a time. The students were fed a liquid formula consisting of 45 percent carbohydrate, 15 percent protein and 40 percent fat. The researchers did not vary the makeup of this formula, which provided about 570 kilocalories per pound.

In one phase of the study, subjects tried to gain weight by ingesting a greater volume of the formula for a long period of time. During another phase of the study, the same subjects tried to lose weight simply by taking in less of the formula. An average size man attempting to maintain his weight would have to consume about five pounds of the formula. Subjects trying to gain weight would have to drink more. For example, in order to consume 4,500 kilocalories one would need to drink nearly eight pounds of the formula.

A 1995 report (Leibel et al. 1995) provided results from more than ten years of observations. The studies showed that people can readily gain a small amount of weight by purposely eating a greater volume of food. However, as they gain weight, their metabolism speeds up and they burn more calories than expected based on the measured weight gain. The metabolic rate overcompensates to slow down weight gain. Typically, someone who increases his weight ten percent by overeating will experience a metabolic rate increase about 15 percent more than would be predicted from their body mass. This will slow the rate of weight gain. On the other hand, someone who loses weight by deliberately restricting food intake experiences the opposite effect. A ten percent weight loss causes a disproportionate reduction in the body's metabolic rate, making it harder to lose additional weight.

Leibel concluded that the body is programmed to keep its fat stores and, as a result, its weight at a preferred level. This series of studies helped explain why the conscientious objectors put on a starvation diet by Dr. Keys were so miserable. Some internal regulator was signaling madly to them that their fat stores were depleted and that they were in trouble. Even after they were allowed to eat freely, they were not happy until they had regained all of their lost fat. Simply eating "normally" again and staying at the new reduced weight was not an option. Their bodies kept prodding them to overeat until they had replenished the fat stores to the level their internal regulator told them was required.

Dr. Sims' attempts to fatten up the Vermont prisoners met the same built-in resistance. Their bodies sensed that they were carrying more fat than they were "entitled to" and kept telling them to cease eating so much and get rid of it. The message is pretty clear. The body does not appear to regulate food intake per se, but instead strives to maintain a certain level of body fat.

Although the control mechanisms for weight regulation are not fully understood, the thyroid gland is believed to play a role. This gland releases a hormone which stimulates cellular metabolism. If you suddenly begin overeating, your thyroid will respond by firing up your metabolic furnace. However, if you suddenly begin eating much less than normal and continue doing so for a period of time, the thyroid slows metabolism down in an effort to conserve vital fat stores. Other feedback loops also are involved. The body is on "cruise control." It is constantly assessing the nature of the diet, the habitual level of activity and the amount of stored fat, making adjustments as needed to maintain weight at a more or less constant level. This automatic process results in many people maintaining the same body weight for long periods of time. We all can remember relatives or friends who seemed to stay the same year after year, always wearing the same "Sunday" clothes.

The groundbreaking studies led by Keys, Sims, and Leibel all strengthen the idea that the body has a set point, a level of fat that is defended automatically. The term set point might be somewhat inaccurate. The body probably does not defend a precise level of fat; more likely, it is programmed to maintain fat levels within a narrow range. Bennett and Gurin, in a 1982 book, present the most complete discussion of set point theory. They describe set point, very simply, as the weight at which you stabilize when you go about your normal activities and do not really think about how much you are eating.

How fat or thin a person happens to be is no accident. Each body "knows" how much fat it is entitled to and balances food intake and metabolic efficiency to maintain it at that level, give or take a few pounds. When the body senses that it is storing either too much or too little fat, food intake and metabolic rate are adjusted appropriately to regain the status quo. Of course, the conscious human mind is unaware that this is going on. William Bennett presented an excellent summary of how all of this works in a 1995 editorial titled *Beyond Overeating*, published in the New England Journal of Medicine.

Set point is not unique to humans. The same process has been observed in a number of animals and probably is universal, at least in warm-blooded species. Researcher Barbara Hansen kept monkeys at a reduced weight for two years by limiting the amount of food they could eat. She was surprised to observe that, once allowed to eat freely, the monkeys persistently overate until they had regained their starting weight of two years before. Although starved for 24 months, their bodies had not "forgotten" what their proper weight should be (Vogel 1999).

Common ground squirrels are a more striking example. They are programmed to put on weight just before winter to get ready for hibernation. During the rest of the year they maintain lower levels of body fat. Researchers deliberately disrupted this highly orchestrated pattern by interfering with their activity levels, disrupting their sleep and depriving them of food. Amazingly, upon being left alone, the ground squirrels quickly adjusted their food intake to return to their normal body weight for that time of year (Mrosovsky and Sherry 1980).

The Precise Cadets

The set point process actually has been monitored in a few cases. During two weeks in 1953, scientists from Cambridge University recorded every activity of 12 British military cadets throughout the day. They also weighed nearly everything the cadets ate as well as the leftovers on their plates. The cadets recorded all between-meal snacks and the calorie content was estimated. Using a series of tests, the researchers measured the energy used in normal cadet activities, such as marching, standing, sitting, and exercising. The cadets then kept a detailed record of everything they did and the time spent doing it on a specially designed form. In this highly regimented setting, the researchers were able to get an accurate measure of how many calories the cadets took in each day and how many they expended in various activities.

Over a two-week period, the cadets ate almost exactly the amount of food necessary to provide the energy required for their daily activities. They were able to maintain this balance while giving little or no conscious thought to how much or how often they ate. Interestingly, the cadets did not eat more on the days they were especially active. Instead, if they expended a lot of energy on a given day, they would eat more food about two days later to make up the deficit in energy. The cadets automatically balanced food intake and energy expenditure with fine precision. Over the two-week period, their body weights stayed nearly constant (Edholm et al 1955).

Another example of the precision with which food intake and energy expenditure are balanced comes from a study of the Swedish national cross-country ski team. Four women and four men members of the team were monitored for one week during a pre-season training camp. The athletes, assisted by dieticians, recorded their food intake. Energy output was carefully measured using the doubly labeled water technique. During the seven days of the study the athletes balanced food intake and energy input almost exactly (Sjodin et al. 1994).

Some of the key concepts underlying the modern set point theory originated with Andre Mayer, a highly-regarded French physiologist of the 1930s. This astute observer noted that most people are able to "manage" their food intake and weight with great precision. The word manage was put in quotes in the last sentence because it is not quite the right term. It implies some kind of deliberate conscious control, which is not the case at all. The process does not rely on the conscious mind. Like the natural control of body temperature or breathing, it just happens. Mayer concluded that two systems function to maintain a prescribed level of body fat. The first system operates on a short-term basis. Mayer observed that both animals and humans have the innate tendency to eat about the right amount of food they need each day to keep their weight at a relatively constant level. But there is some slippage. A person does not eat exactly the amount needed each day to maintain stable weight. So there is a long-term correction. Mayer showed that if people overate one day, there was a good chance that they would compensate for it by eating less on one or more subsequent days. Over a period of time, such as two weeks or a month, body weight stayed essentially the same. Most people in the world manage to maintain the same weight for years without ever thinking about it (Mayer 1968).

Around 1930, some scientists performed what many would consider a bizarre experiment. First they ground up some dog intestines. They then injected a liquid extract from the ground-up intestines into some rabbits. It is hard to imagine what compelled them to do such a thing. In any event, these injections caused the rabbits' gastric secretions and their appetites to decline noticeably. Analytical procedures were crude at the time, so the researchers could not isolate or identify the active agent (Kosaka and Lim 1930).

Nearly a decade later another researcher repeated this study, making some additional observations. He noted that rabbits injected with the dog intestine extract responded by eating less. He concluded that something released by the intestines was reducing normal appetite (MacLagan 1937). More than 20 years later, in the 1960s, some scientists conducted similar studies with mice and rats. They also found that the injections caused animals to eat less (Glick and Mayer 1968).

The causative agent, a hormone, was identified in 1973 and given the name *cholecystokinin*. This difficult-to-pronounce word comes from the Greek terms for gallbladder, sac, and stimulation. A literal translation is something like "stimulates the gall bladder sac." It is easier not to use the word at all, but to employ its common acronym, CCK. CCK has a number of roles in food digestion. For example, it signals the gallbladder and pancreas to release food-digesting enzymes at the correct time. It also prevents the stomach from releasing its contents into the small intestine too quickly (Gibbs et al. 1973).

The 1973 study showed for the first time that CCK released during a meal tells us that we are getting full and it is time to stop eating. Rats injected with CCK ate only half as much at one feeding as non-injected rats. The researchers went even further by "tricking" the rats. Some were treated surgically so that any food they ate immediately drained out of their stomachs and out of their bodies, never reaching the intestines. The rats never received a message from their intestines that food had arrived, so they continued to eat. The researchers then proceeded with step two. They were able to "convince" these poor creatures that they were actually getting food by injecting CCK into their abdomens. Although food was drained from the stomach as soon as they ate it, injecting CCK made the rats think the food was being digested. They stopped eating, groomed themselves and took a nap.

CCK has been termed a satiety factor because it suppresses appetite. Other chemical substances have the opposite effect. For example, a substance called *ghrelin* has been dubbed the "hunger hormone," because it stimulates eating. CCK and ghrelin are just two examples of numerous chemical agents believed to affect food intake, some by stimulating appetite and others by suppressing it. Some scientists have speculated that obesity might be controlled by learning to manipulate the production of substances such as CCK and ghrelin. Others caution that such control is unlikely, arguing that the storage and burning of body fat is too complex and is under the control of too many factors (Kojima et al. 1999).

Andre Mayer was correct in concluding that the body has both a short and long term weight-regulation system. The hormone leptin plays a major role in this. Released by fat cells, it communicates with the hypothalamus, a mysterious organ located deep within the brain. The hypothalamus is about the size and shape of an olive. This small organ plays key roles in regulating many critical body functions of which we are not consciously aware, such as heart rate, blood pressure, and body temperature. It also appears to play a major role in regulating the amount of fat stored by the body.

Scientists observed as early as 1840 that damage to certain parts of the brain could cause eating to go completely out of control and result in enormous weight gain. However, it was not until a century later that the hypothalamus was identified as the center of appetite and weight control. Experiments conducted shortly before World War II demonstrated graphically that surgically damaging the hypothalamus of young rats caused the animals to eat voraciously and become enormously fat. Researchers probably were amazed at what they saw. As soon as the rats awoke, while still dizzy from the anesthetic, they staggered to the food trough and began eating uncontrollably, stopping only when they could hold no more. In one study, several rats were reported to have inhaled food particles until they could hardly breathe, while one rat died, apparently asphyxiating itself with food (Pool 2001).

A damaged or diseased hypothalamus can have the same effect on humans. In 1964 a woman was admitted to a hospital in New York City exhibiting very bizarre behavior. She was constantly hungry and was subject to unprovoked crying or laughing and to fits of uncontrollable rage. If not fed constantly, she would bite, scratch, and throw things at hospital personnel. Doctors eventually found that allowing her to eat continuously was the only way to keep her calm. As a result, she gained more than 50 pounds during her two-month stay in the hospital. She then died of an apparent heart attack at 22 years of age. An autopsy revealed that a tumor had destroyed a large part of her hypothalamus. This damage short-circuited the system that controlled her eating. This caused her to experience constant voracious hunger, no matter how much food she ate — 8,000 to 10,000 kilocalories per day, or how fat she became (Reeves and Plum 1999).

Although scientists knew that the hypothalamus was involved in the regulation of body weight, the underlying mechanism was unknown until the early 1990s. A landmark 1994 paper in the prestigious journal *Nature* announced the discovery of a gene controlling production of the hormone leptin (Zhang et al. 1994). Leptin gets its name from *leptos*, the Greek word for thin. Released by fat, it circulates in the blood and directly controls how much we eat. Too little leptin signals the hypothalamus that not enough fat is being stored, that the individual has fallen below the set point. The body responds by slowing down metabolism and increasing appetite until the missing fat is restored. If someone gets too fat — above the set point, the level of circulating leptin increases to reflect this. Sensing excess leptin, the body speeds up metabolism and decreases appetite until the appropriate amount of fat is shed.

If this balancing act fails, as sometimes happens, the results can be tragic. As described earlier, this can happen if the hypothalamus is damaged by accident or disease. A genetic defect also can be the cause. In 1997, Turkish doctor Metin Ozata examined one of the fattest humans he had ever seen. This 22-year-old man was identified only as Patient 24. He was five feet, six inches tall and weighed 330 pounds, with a 55-inch waist. His appearance was striking in other ways as well. He did not seem to have gone through puberty. His genitalia resembled that of a small boy, and he had no facial or body hair.

Dr. Ozata and a French collaborator eventually determined the cause of Patient 24's problems. Because of a mutation, he was unable to produce leptin. Lacking leptin, his body could not regulate its fat levels. Sensing no leptin in the blood, the young man's body "assumed" that he was starving. The brain sent out frantic signals, "Eat, eat, eat!" Not surprisingly, this poor man was described in clinical terms as "markedly hyperphagic" — which simply means that he ate enormous amounts of food. But no matter how much he ate or how fat he became, he remained in a constant state of starvation. Patient 24 is not unique. Researchers have identified a number of people with the same kind of mutation, including some of Patient 24's close relatives. Nearly all of these cases are in families that are highly inbred, a condition that is still common in many parts of the world. Patient 24's parents were first cousins (Strobel et al. 1998).

A Flexible Set Point

Fortunately, aberrations such as those described above are rare. For most people on Earth, the set point process works almost flawlessly. It keeps body weight within a fairly narrow range for years at a time. However, the set point rarely remains constant throughout life. It can be moved up or down by major lifestyle changes. At any given time, two external factors interact to determine set point. The first is the nature of the diet one habitually eats — specifically, how rich it is in calories and how savory and appetizing it is. The second is the habitual level of physical activity. Two extremes will illustrate this.

A peasant living on a rice and vegetable based diet containing less than 15 percent fat and performing hard labor every day will remain thin throughout life. In contrast, a New York City stockbroker who sits at a desk or a computer for long hours and consumes fat-laden, calorie-rich foods at every meal will stabilize at a much higher weight; most such individuals will be overweight or obese. Drugs are a third external factor that can come into play. It is well known that nicotine from habitual smoking will lower body weight, as will certain commercial drugs (Bennett 1995).

Exactly how the set point might change at intervals during someone's life can be demonstrated using a fictional person, whom we shall call Joe. When we first meet him, Joe is a senior in high school. He is very active and plays on the high school tennis team. His mother shuns fast food and is a big believer in wholesome family meals. Joe nearly always eats breakfast and dinner at home and lunch in the school cafeteria. At the beginning of his senior year, he is five feet, nine inches tall and weighs 160 pounds. During the next year he is weighed periodically. Although he is not on a diet and does not restrict his eating in any way, his weight fluctuates only within a narrow range, from 156 to 164 pounds. Joe is staying at his set point. During his last year of high school he does not alter the energy density of his diet or his habitual level of physical activity. As a result his weight is very stable, varying only a few pounds one way or another.

Upon completing high school, Joe goes away to college and lives in a dormitory. He no longer plays tennis regularly and, since physical education is optional at his college, he does not take it. His main exercise is walking to and from classes. He eats nearly all his meals in the dormitory cafeteria, where he consumes a larger quantity of fatty foods and sugary desserts than had been possible under his mother's watchful eye. He also joins his friends for a lot of late night pizza and beer parties. In addition, he keeps cookies and candy in his room for snacks.

When he returns home for Christmas, his family is surprised to see that he has gained considerable weight, now topping the scales at 185 pounds. Joe remains at college for four years. During that time his weight remains relatively constant, hovering around 185 pounds. The set point process is working. Upon entering college Joe increased the caloric density of his habitual diet. In addition, he reduced his general level of physical activity. His body responded by raising his set point, in effect "defending" a higher level of body fat.

To be clear, Joe could not have gained much weight in college simply by eating more of the same foods he had eaten while at home. In order to increase the amount of fat stored, he had to begin eating a more calorie-rich diet. In a relatively short time, this richer fuel mix raised his body's set point, causing him to stabilize at a higher body weight with more fat in storage. He also became more sedentary in college, and lowering his habitual level of physical activity also signaled his body to store more fat.

After college Joe joins the Army. The physical demands of basic training cause him to shed about 15 pounds. He then volunteers for airborne training and goes to jump school. This highly strenuous regimen results in Joe losing an additional ten pounds, bringing him back to his old high school weight of 160. He remains in the Army three years. During that time he does not vary his habitual level of physical activity or the nature of his diet. As a result, he remains at the newly-established set point throughout his time in the military.

Joe marries soon after leaving the Army and goes to work for an accounting firm. His wife, an excellent cook, treats him to delicious, high-calorie meals. He and his wife also eat out a lot. Joe works long hours at his desk and has little interest in or time for exercise. By the time he has been married two years, Joe weighs 195 pounds. He then stays at about that weight, his new set point, for the next 15 years.

Note that Joe's weight changed only after a lifestyle alteration modified the energy density of his diet and/or his habitual activity level. He then stayed at that new set point as long as his eating and exercise habits remained the same. What if, during the period when he was maintaining a weight of 195 pounds, he had gone on a diet? With sufficient willpower, Joe might have restricted his food intake long enough to lose a noticeable amount of weight. However, as soon as he stopped dieting and started eating "normally" again, the set point would take charge. He soon would gain the weight back and stabilize again at somewhere around 195 pounds.

During his high school years, Joe was storing enough body fat to meet his energy needs for a number of weeks. You might think this fat was stored as amorphous, greasy slabs of tissue coating Joe's chest, ribs, belly and buttocks. However, fat storage in the human body is much more elegant and organized than that. Joe actually had about 30 billion fat cells, each of them containing a clear, glistening droplet of fat. You might think of these cells as 30 billion tiny storage tanks. They have been described as looking like soap bubbles, transparent and roughly spherical, but slightly distorted by the pressure of adjacent cells. Others have described them as looking much like the clear bubble wrap used when shipping fragile items.

Joe changed the size of his fat cells several times during his life. When he went to college and raised his set point, his fat cells began to expand. When he later entered the Army and lowered the set point, his fat cells got rid of some fat and became smaller. Joe's fat cells could have expanded, if required, to about four times their normal size. After that, any additional weight gain would require that Joe generate new fat cells, which his body would gladly have done.

An enterprising scientist named Jules Hirsch came up with a way of counting and measuring fat cells in the late 1950s. He began by extracting a sample of fat tissue from the body and weighing it. He then dissolved the sample and isolated the fat from the other tissues. He collected the fat as individual cells floating on top of a saline solution. The saline solution was then passed through an electronic device that automatically tallied the number of fat cells larger than a set size. By resetting the minimum size with each run, Hirsch got an accurate reading of how many fat cells were in a given sample and what their size distribution was.

He analyzed fat samples from some very obese patients at the Rockefeller University obesity clinic. He found that not only did very obese people have large, well-stocked fat cells; they also had many more fat cells than people of normal weight. The same was true in rats. To become truly obese, it was necessary for an individual, be he rat or man, to greatly increase the number of fat cells. To be more precise, researchers have shown that the average person can become about 60 pounds overweight just by filling up existing fat cells. After that, the body must generate new fat cells for any additional storage (Hirsch and Gallian 1968).

As long as the nature of the diet and the habitual level of activity remain the same, body weight stays relatively constant, varying up and down by only a few pounds. This takes place without conscious thought or any deliberate action on our part. The body constantly monitors the amount of fat in storage. When the actual level varies significantly from the internal standard, an alarm goes off. The body then increases or decreases the amount of stored fat as required to correct the discrepancy.

The precision of this process is amazing. People who do not change their lifestyles often stay at about the same weight for years. If this system were not in place, think about what it would take to gain three pounds a year — 30 pounds in a decade. The average adult consumes about one million kilocalories annually, a rather amazing figure. Theoretically, taking in 10,500 excess kilocalories during a given year would result in the net storage of three pounds of fat. A miscalculation of this size when one is consuming nearly a million kilocalories in the course of a year represents an error of only about one percent. However, as long as the energy density of the diet and the habitual level of activity remain the same, not even an error this small takes place. As expressed by Pennington (1953), "In the obese person of constant weight, as in the lean, the appetite is balanced to the energy output with fine precision...Caloric evaluations cannot match, nor conscious willpower rival, the exactness and persistence of this biological adjustment."

The fact that human weight is maintained with such precision is a serious challenge to the thrifty gene theory, which maintains that, when confronted with abundant food, we will eat "voraciously" and store the excess calories "with extreme efficiency" (Brownell and Horgen 2004). There is convincing evidence that for most people this simply does not happen. Dr. Sims' experiment at the Vermont State Prison demonstrated that it does not. In fact, a main function of the set point mechanism is to prevent it from happening. The last chapter challenged the theoretical basis of the thrifty gene theory. The idea of widespread famine in prehistory, a basic underpinning of the entire concept, was shown to be extremely doubtful. This chapter has cited research to show that the theory also fails to accurately predict events in the real world. People do not gain weight easily when overfed. Instead the body actively resists gaining weight from ever larger amounts of the same foods because of a precise set point mechanism.

Sims had trouble fattening up his prison subjects because he simply fed them more of the same food; he did not change the caloric content of their diets. He could easily have fattened them up by feeding them a richer fuel mix. If he had replaced their normal prison rations with a diet consisting of candy bars, glazed donuts, fast food hamburgers, potato chips, and other calorie-rich foods, and then allowed them to eat all they wanted, nearly all of the prisoners soon would have become much fatter. They then would have stabilized at these higher weights — their new set points, but only as long as they consumed the new calorie-rich diet. If, after a time, Sims had switched them back to their old prison diet, they soon would have shed the extra pounds and returned to their original set points.

Much of the world is in the throes of an obesity epidemic, along with diabetes and the other metabolic diseases that accompany it. This is a singularly modern problem. Very few humans, if any, became obese or diabetic during the time we lived as hunter-gatherers or as peasants. The physical demands of day-to-day life and the low energy density of the foods available would have prevented it. Obesity and diabetes accompanied the advent of agriculture, but only a few members of the ruling classes were affected.

Europeans began to recognize obesity as a worrisome problem about 300 years ago. It became widespread only during the last century and has begun to reach epidemic proportions only during the last 40 years or so. This happened because people in America and other industrialized societies began to replace traditional foods with an array of enticing, high-calorie industrial foods. The effects of this much richer fuel mix soon became evident — the human race began to become obese at an alarming race. Even more alarming, as our species became fatter, we began to experience increasing levels of diabetes, hypertension, strokes, and heart disease.

In order to fully understand this modern predicament, we need to go back to the very beginning, when humans first evolved. During the last century, our species began to live and work in air-conditioned buildings, to eat processed, pre-packaged foods, and to drive cars nearly everywhere we go. But beneath our skin and deep within our cells we still are creatures of the forests, plains, and savannas. Living off the land as hunter-gatherers for most of our time on Earth shaped our bodies and metabolic systems. Accordingly, the next chapter takes a close look at how the first humans evolved, how they lived, and the kinds of food they ate. This will help us to understand why the foods we now consume are making us so obese and so sick.

Chapter 4. The East Side Story

Modern humans, *Homo sapiens*, first appeared between 150,000 and 200,000 years ago, but events that shaped our diet and our bodies began to take place much earlier. These events played out in Africa, but in an Africa much different from today. Ten million years ago a rich, nearly unbroken forest covered most of the continent from east to west. This primeval expanse of rich jungle, which had persisted virtually unchanged for millions of years, was home to numerous species of apes. These early primates had flourished in this benign, nurturing climate for eons, but things were about to change for the worse. The continent was growing restless and momentous forces were beginning to stir beneath the surface. Soon the earth shuddered and began to rip itself apart, thrusting up large chunks of what is now Ethiopia and Kenya, forming mountains nearly two miles in height. Areas in the rain shadows of newly-risen mountains became drier. The once continuous expanse of well-watered jungle began to fragment, changing into a mosaic of open forest, dry savanna, and shrubland.

Additional changes were in the offing. Further contortions of the earth created a spectacular gorge in east Africa, running north to south for hundreds of miles. This geologic marvel, the Great Rift Valley, created a barrier which stranded apes on the east side of the Valley in an increasingly dry and challenging environment. Yves Coppens, a noted French anthropologist, argued that formation of the Great Rift Valley was the crucial event launching our early ape ancestors on their long journey to becoming human. According to Coppens, the common ancestors of modern humans and chimpanzees were suddenly separated into two isolated populations occupying contrasting environments. Unable to interact or inter-breed, they went their separate evolutionary ways. The western descendants of our common ape ancestors continued to evolve, but they evolved in the forest, with sheltering trees always near.

On the other side of the Great Rift Valley, the eastern descendants of these same common ancestors took a different path. Forced gradually out of the trees, they faced a daunting evolutionary challenge — learning to live on the ground and in the open, beyond the shelter and safety of the forest. Adapting to this dangerous, challenging environment transformed both their bodies and minds. It took many generations and many failures, but they eventually mastered this new environment and in the process began to evolve into human form. Coppens (1994) has referred to this sequence of events as the "East Side Story."

East Africa became progressively drier over time and food plants became more widely dispersed. The apes marooned there were forced to spend increasing amounts of time on the ground traveling between food sources. Ultimately, this favored bipedalism — walking on two legs. Over time, natural selection favored those in each generation whose rear paws had become more adapted to walking on the ground. This began the long evolutionary process eventually leading to the human foot. Front paws began to change as well. No longer needed to support the weight of the body when walking, they increasingly were used to manipulate simple tools or to wield weapons. In a few million years, ape paws would evolve into human hands, capable of almost unbelievable dexterity and sensitivity. Charles Darwin (1871) recognized the survival advantages of walking upright and having the hands free more than 100 years ago.

> If it be an advantage to man to have his hands and arms free and to stand firmly on his feet...then I can see no reason why it should not have been more advantageous to the progenitors of man to have become more erect or bipedal. The hands and arms could hardly have become perfect enough to have manufactured weapons, or to have hurled stones and spears with true aim, as long as they were habitually used for supporting the whole weight of the body...or so long as they were especially fitted for climbing trees.

These early ground-dwelling ancestors of ours are referred to as *hominids*. Survival required them to adopt a life that involved traveling increasing distances on the ground. Despite this, for a long time they remained similar to their tree-dwelling cousins in many ways. They had the same small brains, large cheek teeth, and protruding jaws. For four to five million years the hominids continued to live much as apes did, consuming mostly plant foods. It is believed that their social organization was similar to that of modern baboons. But these humble creatures were taking evolutionary steps that eventually would lead to modern humans.

In addition to gradually perfecting the foot and hand, they set another process in motion that would revolutionize the world. A geological accident had marooned them in a challenging, dangerous, deteriorating environment. They increasingly had to rely on superior intelligence for survival, a strategy that accelerated the trend toward the development of a larger, more complex brain. The early acceleration in cognitive ability, along with development of the hand and the foot, set the stage for the eventual appearance of modern humans.

Food and Intelligence

Huge brains are the key to human dominance of Earth, but the survival value of intelligence did not come to the fore recently nor did it originate with humans. Instead, increasing encephalization, as biologists refer to the process, has been going on throughout evolution. For example, the first primitive mammals, appearing about 230 million years ago, had brains four to five times larger than the reptiles with which they then shared the world. Millions of years later, primates evolved brains twice as large as other mammals. Apes continued the process, evolving brains twice as big as other primates. Humans carried the trend to an even greater extreme, evolving brains three times larger than those of apes.

What caused the human brain to grow so massive? Scientists now believe that our ape ancestors became increasingly smarter because survival in a more dangerous, more complex environment required it. This has been termed "ecological intelligence." Researcher Katherine Milton (1981) observed a striking example of this in the jungles of Panama. For three years she followed troops of monkeys through the humid rain forest from dawn to dusk. The two species she observed are roughly the same size, but have radically different lifestyles. The placid, lethargic howler monkeys eat mostly leaves, have a long intestinal system, and take as long as 20 hours to digest a meal. Since they can acquire their food from a relatively small area, they do not range very far when feeding.

In contrast, the active, energetic spider monkeys are primarily fruit eaters. They have much shorter intestines and can digest their food in about four hours. Spider monkeys might have to range over several thousand acres to meet their food needs. In addition, they have to learn the ripening times of 50 to 100 different fruits. Milton found striking differences in the brains of the two monkeys. The more-active and wider-ranging spider monkeys have brains more than twice as large as the howlers. They also take longer than the howler monkeys to grow up, reflecting the additional time needed for youngsters to grow a larger brain and to master a more complex food environment.

Milton and others have uncovered a general principle of great importance: Animal species with diverse diets consisting of foods widely dispersed in a complex environment develop greater intelligence. Larger brains are needed to store and process increased amounts of ecological information. The survival value of greater ecological intelligence was the initial impetus to increasing brain size during early hominid evolution. Prehuman species that managed to survive and reproduce in the open woodlands and savannas did so to a great degree by becoming cleverer. Surviving in a complex mosaic of forest and grassland put a premium on learning and remembering the location of food sources scattered throughout a large area. This need to store and process information started a long trend in evolution, leading inexorably to upright species with increasingly larger brains and greater cognitive abilities.

The land east of the Great Rift Valley continued to get drier, until eventually savanna and scattered forest covered most of it. Food-bearing plants became even more widely dispersed. As a result, hominids were forced to range over increasingly larger areas, often many square miles, to find sufficient food. In addition to becoming more mobile, they were compelled to eat a more diverse diet. They learned to harvest a wider array of foods, including increasing quantities of bird eggs, small mammals, turtles, insects and other prey they could easily catch and kill. This required them to learn even more about their environment. Brain size continued to increase during hominid evolution.

Our hominid ancestors inhabited the savannas of east Africa for several million years. During that long time period, many species evolved, flourished for a time, and then died out. Finally, about two million years ago, an especially clever species appeared. They had developed a revolutionary skill, the ability to fashion certain stones such as obsidian and flint into tools and weapons with cutting edges. Their tools and weapons were very crude at first, little more than sharp-edged fragments fashioned by breaking rocks apart almost at random. But as time went on they developed greater skill. By about 1.5 million years ago, they were producing finely-made weapons and tools that were symmetrical, with finely-flaked edges.

In addition to being pleasing to the eye, their products worked very well, enabling them to remove the tough skin from large animals and to sever muscle from bone. Stone tools also were effective in breaking open the skull and long bones of animals, providing access to the rich stores of fat in the brain and bone marrow. With sharp-edged stone weapons and tools, these hominids were able to kill, skin and butcher large prey for the first time in history. This brought about a dietary revolution — one which was to have enormous impacts on human culture and biology.

Skinning and butchering a large animal with a piece of rock, no matter how skillfully fashioned, might seem improbable. But several modern stone workers have shown that it can be done quite effectively. Don Crabtree is a well-known example. He perfected his skills by fashioning stone blades, scrapers and projectile points for many years as a specialist in American Indian artifacts at the Ohio State Museum. He later held a similar position in Idaho.

In 1966, Crabtree was visiting an archaeological site in Arizona to demonstrate stone-working techniques to students helping with the project. A bear invaded their camp and had to be killed by the local cattle manager. This was a ready-made chance to show how well stone implements actually function as butchering tools. The test took place the following morning. Crabtree prepared a set of sharp-edged stone tools and the cattle manager, Mr. Seely, though amused and skeptical, proceeded to skin the bear. He was in for a surprise. Upon beginning the first cut, he found himself off balance as he tensed his muscles in anticipation of resistance. But the obsidian blade sliced into the tough skin smoothly and easily. A somewhat astounded Mr. Seely finished the job in less than two hours. He admitted that the same task would have taken him more than three hours if he had used his finely-honed steel blade (Pfeiffer 1969).

Not all kinds of stone can be fashioned into sharp weapons or tools. For example, pieces of sandstone, shale or coarse granite would not work at all. But certain rocks, such as obsidian and flint, will form extremely sharp edges when carefully flaked or broken. Obsidian was used in the example above. Early humans sought out deposits of such suitable stones and often traded them over large areas. Their trademark product was a finely-made teardrop-shaped hand axe, very standardized in shape and found across much of Africa.

This tool tells us a lot about our early ancestors. First, they were deliberately imposing a shape on the raw material, showing that their minds had an internal model of the tool. Thus the hominid brain was already taking human form nearly two million years ago. Additionally, anthropologists have determined from examining their tools that most early toolmakers were right-handed, similar to modern humans (Toth 1985). This is evidence that the brain might already have been developing a distinct left lobe, an event that would have paved the way for complex language. In fact, these early stone workers probably were capable of rudimentary speech (Laitman 1984). Studies of intact skulls suggest that they had brains weighing about two pounds — twice as large as their ape cousins', but not as large as modern humans'.

Some researchers have estimated that their mental and vocal abilities equaled those of a six or seven-year-old modern child, or perhaps even older. Scientists refer to this early human species as *Homo erectus*, or "upright man." These tough, resilient relatives of ours lived on earth a long time and eventually developed brains falling into the lower size range of modern humans. *Homo erectus* was a clever and resourceful species — well worthy of respect and definitely more human than ape.

Hunting and Eating Meat

"Not in innocence and not in Asia was mankind born." So begins Robert Ardrey's influential 1971 book, *African Genesis*. The phrase "not in innocence" refers to the fact that hominids have been killing and eating other animals for two million years. This phrase also reflects the fact that as our ancestors became self-aware they began to question the morality of killing. A theologian in the 1600s wrote of "Those artless Ages when Mortals lived by plain Nature... Men were not carnivores...and did not feed upon flesh." Much later William Golding, in his novel *The Inheritors*, would describe the mixture of revulsion and pleasure with which a prehistoric man tears into the carcass of an animal he has just killed, savoring its "its rich smell of meat and wickedness" (Pfeiffer 1969).

Some people are so repelled by the idea of eating meat that they refrain from doing so. By exercising discipline, eating the right combination of plant foods and perhaps using nutritional supplements, a few manage to live as exclusive vegetarians or vegans. But there is no record of any successful child-rearing society that did not eat at least some animal products. Trying to rear healthy children on a totally vegan diet is a risky undertaking. Growing children do not necessarily have to eat flesh, but it is best that they consume some kinds of animal foods — such as eggs, milk or cheese — in order to insure healthy development. Despite the repugnance that meat-eating inspires in some, most members of our species will continue to consume animal flesh for the foreseeable future.

Some people object to eating meat on moral grounds, imagining that animals suffer both great physical pain and abject terror while being killed. The following account by David Livingstone (1872), the famous missionary and explorer, might provide some solace. As a young man, he was attacked and nearly killed by a lion. Luckily he escaped, but not before being severely mauled and losing the use of his left arm for the rest of his life. Later he described the event in his journal.

> I saw the lion just in the act of springing upon me. I was upon a little height. He caught my shoulder as he sprang and we both came to the ground together. Growling horribly close to my ear he shook me as a terrier does a rat. The shock produced a stupor similar to that which seems to be felt by a mouse after the first shake of the cat. It caused a sort of dreaminess in which there was no sense of pain nor feeling of terror, though quite conscious of all that was happening...The shake annihilated fear, and allowed no sense of horror in looking round at the beast. The peculiar state is probably produced in all animals killed by the *carnivora* and, if so, is a merciful provision by our benevolent Creator for lessening the pain of death.

Livingstone's fascinating narrative casts doubt on the widespread belief that animals being killed undergo almost unbearable horror and pain. They probably do not. If humans are protected by a kind of anesthetic stupor, it is probable that animals with lesser intellect and self-awareness also have been granted this "merciful provision by our benevolent creator." I have read several accounts of people feeling the same state of surprising calmness when their vehicle goes out of control on an icy road and they feel themselves sliding toward an unavoidable collision.

Although their awakening conscience might have bothered them, the first humans were avid hunters and meat eaters. Many have the idea that our kind started out as gentle vegetarians and only much later began to kill and eat other animals. On the contrary, a lust for meat has been in the primate blood for a long time. It is now known that male chimpanzees regularly conduct organized hunts to acquire meat. Their favorite prey is the red colobus monkey. When chimpanzees make a kill, they usually crack open the skull and eat the brain first. They then shatter the long bones to get at the marrow. Brain and bone marrow contain the highest concentrations of fat in the body — especially in very lean animals living in the wild. There is strong evidence that early human hunters followed these same practices.

Although early humans were not the first members of the ape family to hunt, they were by far the most effective, soon becoming the most skilled predators and killers among land animals. With the advent of stone tools and weapons, they were able to kill animals large and small, and in great numbers. This enabled them to exploit abundant new sources of food — food high in protein and minerals to build larger bodies and high in essential fats necessary to build larger brains. Essential fats are those needed to build human cells and tissue, but cannot be manufactured by our bodies. They are abundant in some green plants, in oily fish, and in the brains and bone marrow of animals. These rich sources of essential fats provided early humans with the biological material required to build the large, intricate brains that soon would give them dominion over Earth.

Some may doubt that humans, even with very sharp stone spears and axes, could kill large, powerful animals. If you are a skeptic, the following description might alter your opinion. Dentist, nutritionist, and world traveler Weston Price was in Uganda during the 1930s. He is the Dr. Price in whose honor the Price-Pottenger Nutrition Foundation is named. He traveled to every inhabited continent and wrote a fascinating book in 1939 describing the diets and dental health of people throughout the world. His book, *Nutrition and Physical Degeneration* (Price 1997), has gone through numerous editions.

While in Uganda, Price observed how pygmies, the smallest of humans, managed to kill elephants, the largest land animal, with surprising ease. Typically working in two-person teams, they first disabled the animal by severing the tendons of both hind legs. This usually took about two days, during which they worked stealthily from behind, taking care that the elephant did not catch sight of them. After immobilizing their prey they attacked openly. While one distracted the animal by shouting and waving his arms, the other moved in and began progressively hacking off the trunk. Gradually the elephant would lose consciousness from loss of blood and eventually die. Pygmies were the only people in the district permitted to kill elephants without a permit.

Scientists have discovered many instances in which "primitive" humans slaughtered large numbers of animals such as buffalo or wild horses at one location. Near Solutre, in France, the remains of an estimated 10,000 wild horses were found. It was clear that early hunters repeatedly used this site to stampede horse herds over a nearby cliff so they would be killed or disabled in the fall. This apparently was a common hunting strategy for thousands of years. The fossil remains of 300 buffalo recently were excavated in Colorado. Ten thousand years ago a group of human hunters drove them down steep, sloping ice into a ravine filled with deep snow drifts. Once immobilized, they were slaughtered and used for food.

Lewis and Clark described Plains Indians using similar techniques in the early 1800s. Mounted Indians drove frightened herds of buffalo over steep bluffs in order to kill them in large numbers (Farb 1968). Although it is not clear how they were killed, the piled-up remains of more than 100 wooly mammoths slain by early hunter-gatherers were discovered in eastern Europe (Wharton 2001).

A Sharing Species

For several years I helped care for my young grandson. During that time, we spent many hours at a city park which had a large, well-shaded sandbox. Various parents had donated shovels, buckets, and other toys for children to play with, but there rarely were enough toys for all of the kids to have one. In addition, there were a few highly coveted toys that several children often would want to play with at the same time. As a result, the young suburban mothers who came to the park would frequently admonish their children to share with others. Not a day went by without at least one child receiving a gentle lecture on the importance of sharing. It was clear that, to these suburban mothers, sharing is a moral imperative and an important requirement for being a good person.

But where did such an imperative originate? There is now strong evidence that sharing emerged as a value very early in our evolution and, like meat eating, was instrumental in making us human. There also is evidence that meat eating and the human values of sharing and fairness have a common origin. In the absence of refrigeration or other means of preservation, meat must be eaten quickly before it spoils. Thus early hunters faced a dilemma — what to do with the horse, deer, pig, or other large animal they had just slain. One person or even a small hunting group could not eat the entire carcass, but any of the meat not consumed would quickly spoil.

Our ancestors came up with a solution that was both simple and inspired. A successful hunter shared part of any kill with members of the hunting group and with other clan members left back at the home base. In return, he could expect a share of future kills made by others. This strategy greatly improved the economic security of the entire group. Over time, groups of hunter-gatherers developed elaborate protocols for sharing meat and learned to keep a precise mental accounting of who was indebted to whom. A successful hunter who shared meat with others earned the right to be repaid in kind at some later time when he himself had a run of bad luck. Hunter-gatherers stored food, not in granaries or larders but in the stomachs of their comrades, who were aware of their obligation to repay.

Meat Sharing and Intelligence

Hunting big game provided large amounts of protein and essential fats, which enabled early humans to increase brain size and complexity over time. These nutrients provided the wherewithal for brain growth, but what actually stimulated it? We have discussed the role of ecological intelligence in early primate brain development. Apes marooned east of the Great Rift Valley responded to their new, more challenging environment in part by becoming smarter. But developing the intellect needed to master a complex natural environment can account for only a limited amount of encephalization — a scientific term for increasing brain size. Some scientists have postulated that early tool-making was a further impetus to brain development, but the role of so-called technological intelligence also would have been limited. A prominent neurobiologist stated in a 1991 lecture that, "It seems to me to be an inadequate explanation, not least because tool-making can be accomplished with very little brain tissue" (Leakey 1994).

Organized group hunting required cooperation, discipline and a measure of intelligence. But, as evidenced by chimpanzees and wolf packs, successfully carrying out such activities does not require an inordinate amount of brain power. The demands of group hunting can account for only a little of the amazing increase in brain size and intelligence experienced by early humans. The most widely-accepted explanation of how humans developed such massive brains came in 1978, when Glenn Isaac published his classic paper, "The Food-sharing Behavior of Protohuman Hominids."

Isaac argued persuasively that it was not hunting itself that had such an influence on the growth of the human brain. Instead, it was the collaboration and sharing of meat in a communal setting that hunting made necessary. In Isaac's view, the adoption of food sharing would have fostered the development of language, complex social interaction, and enhanced intellect. Isaac's theory has been endorsed by many leading anthropologists, including Richard Leakey (1994), who concluded that food-sharing was the most likely candidate for explaining what set early humans on the path to becoming *Homo sapiens*. A willingness to share with others; reciprocity, feeling an obligation to repay a favor or a good deed; and an inborn sense of fairness are some of our most valued human traits. It is chastening to think that all of these tendencies most likely became part of the human character as a result of the killing, butchering, and communal sharing of meat.

It seems amazing that such a seemingly simple process as sharing meat in a social setting could have such far-reaching effects. But the evidence that it did is very convincing. Most anthropologists now believe that the need to function politically and socially in a competitive social environment had a greater influence on brain development than did challenges in dealing with the natural environment. Early humans most capable of exploiting their social environment could form alliances with the most able hunters and gatherers in the band. As a result, they could acquire more food, secure a supportive mate, and leave more offspring. This put a premium on what has been called social or Machiavellian intelligence (Humphrey 1976). With the passage of time, social intelligence and the culture and language that it engendered became increasingly important to human success.

In time, cultural evolution became far more important than biological evolution in enabling human societies to dominate the world around them. This, in turn, stimulated even more encephalization. There seemed to be a runaway arms race between culture and brain growth. Language, technology, social networks, knowledge — all of the elements of culture continuously became more complex. This stimulated more brain growth. Bigger and cleverer brains then drove cultural complexity to new levels. This in turn provoked even greater brain growth, which then led to even more complex culture and so on in a continuously reinforcing cycle (Wills 1993).

The human brain gave our kind immense power over other living beings and over the physical world, but it came at a cost. Typically weighing about three pounds, the brain is an expensive organ both to grow and to keep functioning. It accounts for only two percent of body weight, but uses 20 percent of the body's energy supply. It is so active and so demanding that an amount of blood equal to its own weight must be pumped through it every minute. Humans had to make a number of biological and cultural changes in order to grow and maintain such large brains. First, they had to overcome natural limitations of the female pelvis. Human hips widened as the brain evolved and grew larger, enabling human infants to be born with larger brains and, as a result, larger heads.

Despite these accommodations, human infants still were forced to come into the world much too early. After birth, the human brain must triple in size to become fully functional. In contrast, newborn apes need only double their brain size as they grow up. This means that human babies have a long infancy and a long juvenile period. Human infants are much more helpless than newborn apes, and remain so for a much longer time. In order for a human baby to be born with the same level of physical development as an ape, the human gestation period would have to be 21 months. How many women would want to carry a child for that long and then give birth to a baby nearly the size of a two-year old?

Getting children through a very long infancy and childhood became a major task for our species. There were many hazards to avoid, including venomous snakes, deadly insects, poisonous plants, hungry predators, and the constant danger of fatal falls or drowning. Intelligent, perceptive mothering became of paramount importance. Keeping a child safe — in fact, keeping a child alive, required constant vigilance. An intelligent and watchful mother enhanced the chances that a child would survive to adulthood. But the mother needed help. An attentive male was needed to provide additional food and to protect both child and mother during vulnerable periods.

The demands of child rearing required that humans forge an enduring relationship between the sexes. Most group hunters such as lions, wolves, wild dogs and even chimpanzees are promiscuous, with little long-term bonding between mates. Humans deviated from this pattern, developing high levels of mutual attachment and cooperation within breeding couples. In addition, they invented something unique in the animal world — gender specialization in food acquisition. Increasingly, males concentrated on hunting game while females specialized in gathering edible plant foods.

This arrangement greatly increased food security. Plant foods are a highly reliable source of calories and nutrition. Rarely would women come back to camp empty handed. Because of the variety and quantity of plant foods in most landscapes, some kinds of food would be available throughout the year. Hunting was less reliable, but the high quality protein and fats from periodic kills were highly prized and critical to maintaining a high level of nutrition.

Mastering Fire

According to legend, Prometheus stole fire from the gods on Mount Olympus. The gods, enraged by his audacity, chained him to a rock as punishment. Each day an eagle would appear, tear into his side and start devouring his liver. The liver would grow back overnight, only to have the eagle appear again the next day and start the ordeal all over again. This is a good story, but the way in which our ancestors actually gained dominion over fire might be even more interesting, if we knew it. Hominids did not learn to use fire for a long time, eating raw meat for perhaps a million years.

But wildfires were a common occurrence in their world, since they are so much a part of many natural landscapes. Most are caused by the estimated eight million lightning bolts that impact the earth every day. To put this in more meaningful terms, 100 bolts of lightning strike somewhere on earth during every single second of every day. Even a strike of average power is pretty impressive, discharging enough electricity to power the typical American home for a day and a night. Most last only a few millionths of a second. But about one in ten lasts very long, at least as lightning bolts would reckon time — as much as half a second. It is these super bolts that cause most natural fires. They can generate extremely high temperatures, more than sufficient to start a grass fire or to ignite a standing tree. Lightning causes about 50,000 fires worldwide each year, including an estimated 10,000 in the United States. Tree ring studies of one forest in California showed that a forest fire had occurred about every eight years from the 1600s to the beginning of the 20th century (Page 1983).

Fires occurred frequently on the savannas and in the open forests where early humans lived. They would have feared it at first and marveled at its power and mystery. But being human, they eventually would begin to test it and experiment with it, and eventually harness it. Scientists recently found evidence in a campsite more than a million years old suggesting that a determined human had dragged a burning log into a cave and then kept the fire going. Such opportunistic use of fire occurred early in our history and undoubtedly became widespread, but the daunting achievement of actually making a fire remained thousands of years in the future.

But even if they had to take advantage of chance encounters, early humans recognized the value of fire and began putting it to use. They found many uses for this marvel of nature. It kept them warm during long, cold nights and helped them to keep dangerous predators at bay. In addition, they must have enjoyed basking in its warm, friendly glow, which formed a cozy refuge in an often dark and foreboding world. Soon humans found another use for fire, one that would change both their biology and their culture forever. They learned to cook. This technology greatly increased the overall variety of the human diet and provided new sources of calories and nutrients. James Boswell is one of the earliest to recognize cooking as one of the few activities unique to our species. In his 1773 journal, he defined man as a "Cooking Animal."

Plants such as potatoes, taro, cassava, rice, corn, wheat, and legumes can provide a lot of energy and nutrition, but humans cannot digest these foods efficiently in their natural states. Just imagine trying to eat three or four raw potatoes or several cups of uncooked corn. Doing so would cause digestive upsets in most people. The starch in grains, tubers and legumes is composed mostly of long chains of carbon atoms. As the wheat, bean, corn or other crop ripens, the starch molecules become tightly wound like a ball of twine, forming granules. Humans have difficulty eating these granules when uncooked, but upon heating they burst open and become soft, more palatable, and easier to digest.

The judicious use of fire suddenly made a lot of high energy vegetable foods available to early human populations. Cooking also made meat more calorie-rich. A given weight of cooked meat typically provides 50 percent more useable calories than the same weight of raw meat — which humans ate for much of our history. Cooking, along with food sharing and gender specialization, greatly increased food security. Hunter-gatherers were slowly but surely expanding their food supply through the use of cultural adaptation and technology. This trend would continue and eventually accelerate.

Eating cooked food began to reshape the contours of the human head and face. Chewing tender meat and soft, cooked vegetables could be accomplished with less massive jaw muscles and smaller teeth. Both diminished in size with time, leading to changes in the rest of the face and skull as well. Massive brow ridges and other thick bony protuberances had evolved to anchor large, powerful jaw muscles. These bones became less massive as the jaw muscles began to shrink in size. The skull itself was able to become less massive and thick, making even more room for an ever-growing brain.

Modern Humans

Sometime between 150,000 and 200,000 years ago, a new kind of hominid evolved, an especially clever, predatory species of ape that in time would have the temerity to refer to itself as *Homo sapiens* or "wise man." *Homo sapiens* would be the last and, in time, the sole remaining hominid species. Despite difficulties and setbacks, our human ancestors thrived, soon spreading throughout all of Africa. Sometime prior to about 60,000 years ago they spread out of Africa and began to settle the Middle East. About 45,000 years ago another migration began. Some early humans headed north and east toward the Indian subcontinent and Asia. Others headed north and west, eventually settling Europe and what would eventually become the British Isles. By about 15,000 years ago, *Homo sapiens* had spread to every habitable continent and were the only hominid species left on earth.

The fact that there is only one species of hominid on Earth requires some explanation. This is unusual in the animal kingdom. For example, worldwide there are numerous species of large cats. The same is true of many other animals — such as squirrels, foxes, rabbits, bears, deer, and monkeys. The list could go on and on. Why are we represented by only one species? It was not always this way. When *Homo sapiens* first appeared on Earth, they shared it with other hominid species. But these others did not last long. As our intelligent and ruthless ancestors became more numerous and more powerful, they probably hounded these poor creatures to extinction. A few of the odd-looking people who shared the earth with us for a while left bones and artifacts behind, but most persist only in our racial memory as legends and myths.

Nearly every culture has told stories of strange beings, misshapen but somehow human-like that once lived among us. We in the English-speaking world have long frightened and fascinated our children with tales of dwarves, giants, and trolls. It is unlikely that these "mythical" races are totally fictional. Creatures that inspired such stories probably coexisted with humans for a time, but not for long. These poor creatures occupied an untenable position — recognizable as like us but still so different as to inspire fear and revulsion. A line from an early horror movie comes to mind, "One should not suffer a monster to live!" It perhaps is best that these closest relatives of ours did not survive; they would have been degraded and abused in our modern world.

What accounts for the astounding success of early humans in moving out of their tropical homeland and occupying every part of the Earth, from equatorial jungles and savannas to harsh deserts and north to the frigid arctic? One of our strengths might be surprising, since we in the modern world have come to think of ourselves as weak and puny specimens, no match for many of the strong, fierce creatures that still prowl the natural world. This is far from reality. Humans actually are quite formidable physically when armed with even the most primitive weapons. This is especially true when we organize ourselves into small hunting bands or fighting units (Ardrey 1976).

Humans are unique in the breadth of our abilities. Of all living things, we are the only species capable of running 20 miles during the heat of the day, then descending a steep cliff to a river, swimming the river, ascending a cliff on the other side, then crawling for a mile through dense jungle, and finally, climbing a tree after setting a trap to catch an animal on the trail below. After capturing the animal, we are the only species capable of weighing alternatives — of deciding, in the words of Peter Farb (1978), whether to "ride it, paint a picture of it, worship it, or eat it."

Being without fur and having sweat glands gave early humans a big advantage over other animals, especially on the tropical savanna. They could run for long periods of time without overheating, even during the hottest part of the day. This enabled them to move freely and hunt when other predators were inactive and prey animals were sluggish. In a hot climate, humans have the ability to run almost any other animal to exhaustion. Humans have other physical abilities that helped them become the most efficient hunters on Earth. One of these is handedness, being preferentially right or left handed. Handedness enables humans to throw rocks, spears, or other objects with great force and accuracy. For example, baseball pitchers regularly throw balls at speeds in excess of 100 miles per hour.

Based on their rugged skeletons, Paleolithic hunters were a match for many modern athletes. They had other physical advantages. They had inherited depth perception and stereoscopic vision from their tree-dwelling ancestors. Color vision was another valuable endowment from their primate relatives. These attributes, developed while living in the trees, were equally valuable on the ground. The ability to judge distance accurately and to distinguish subtle differences in color enabled early humans to see other animals clearly against a background of vegetation. These capacities were crucial both in hunting and in avoiding attack by large, dangerous predators.

Physical prowess did much to ensure that our kind not only survived, but actually prospered in a dangerous, challenging world. But our biggest advantage was an unparalleled ability to learn, to reason, and to plan. Phenomenally large brains enabled hunter-gatherers to store and retrieve large amounts of information about their environment. It also enabled them to analyze the world around them and exploit this knowledge to enhance their chances of survival. Such reasoning and analytical ability enabled humans to fully exploit any environment and to outsmart any other animal on Earth. Some have described the human brain as an evolutionary Swiss army knife, since it can be put to use in so many ways. This analogy is a very apt one, since evolution has demonstrated that the powerful human brain is the supreme all-purpose tool.

Modern anthropologists closely studied a number of hunter-gatherer groups that survived into the twentieth century. They were amazed at the voluminous, detailed environmental information that these "primitive" people carried around in their heads. Hunter-gatherers, regardless of where they lived, had an almost encyclopedic knowledge of the plants and animals in their area. For example, one observer noted that the Bushmen of southern Africa were intimately acquainted with the life habits of more than 50 animals. She further observed that Bushmen knew the location of nearly every stone and bush in their territory and could take a visitor to every site of edible plants, fruits, or nuts and describe the time of the year in which the foods would be available. They usually had named such places, even if they were extremely small (Thomas 1959).

Intelligence and ingenuity have enabled humans to procure a surprisingly good living in even the most forbidding terrain. The Inuit of the far north are a good example. You might imagine that, prior to contact with the "civilized" world, they subsisted on a monotonous diet of seal blubber and fatty fish. This is not the case. Dr. Weston Price visited a group of them in the 1930s and observed that their diet was surprisingly varied. At the time of his visit, their available foods included caribou, quantities of frozen fish, and the flesh and organs of large sea mammals. They also had quantities of ground nuts which had been pilfered from the underground caches of field mice and kept for the winter. Their meat and fat based diet was supplemented by kelp and several varieties of frozen berries (including cranberries) which had been gathered during the summer and frozen for winter use. Their diet also included the blossoms of flowers and sorrel grass which had been preserved in seal oil (Price 1994). For a people living in a supposedly frozen, barren landscape, the Inuit were living quite well.

Once learning the habits of animals, early humans excelled at applying this knowledge to efficiently capture or kill them. Intelligence and cunning turned our species into perhaps the best stalkers and hunters on Earth. We already have discussed how bands of humans would stampede frightened horses, buffalo, and other herd animals over cliffs, easily killing animals that were large and fast, but not very intelligent. Following are a few more examples of human ingenuity.

The natives of some south sea islands devised a nearly fail-proof way to capture a large species of land crab. Reaching the weight of several pounds, the crabs made very good eating. They commonly climbed coconut trees after dark to reach the fruit. Throughout the night the industrious crabs cut off the coconuts and let them fall to the ground. Near morning, when their harvest was complete, they made their way back down the tree rear-first. If they were unlucky, they might encounter a girdle of grass the clever natives had placed around the tree about 20 feet above the ground. When their rear claws touched the grass, the crabs thought they were on solid earth and let go. The long drop stunned them and they were quickly scooped up by the islanders waiting below.

Many animals make easy prey because their escape tactics are totally instinctive or "hard-wired." For example, a tribe in Malawi learned to catch a kind of partridge called the spur fowl because they could predict what the bird would do when disturbed. When first frightened, the bird took off and flew about 100 yards. The second time it was alarmed, it flew about half that distance. If disturbed a third time, it flew only a short distance. There it remained hunkered down until the human hunter simply approached and picked it up (Price 1994). Numerous examples such as these could be cited from around the world. Wherever they lived, human hunter-gatherers studied the habits of the animals they used for food and devised clever, efficient ways of capturing or killing them.

The human brain is so effective because it enables us to learn and store large amounts of information, to solve problems in the abstract, and to develop and implement complex strategies. It also gives humans the unique ability to invent things — to create tools, weapons, forms of shelter, or art that did not exist in the world until they were conceived by a human mind. People differ in their ability to do these things. A small number of people, perhaps less than three out of every 100, possess exceptionally powerful intellects. Such rare and important individuals are sometimes said to possess genius, which Samuel Johnson defined as "a mind of large general powers accidentally determined to some particular direction."

In modern terms, this would include people with IQs of about 130 or above who have trained their minds to function well in a particular area. We in the modern world are highly dependent upon such individuals to ensure that our complex societies continue to function. This cognitive elite also is responsible for most of the technological and cultural innovations that make progress possible. The same was true more than a thousand centuries ago. Then as now, a select few individuals blessed with unusually perceptive and powerful minds led the way in developing new insights and making new discoveries. These clever innovators were instrumental in enabling early humans to master their environment and occupy the entire world in a relatively short time.

By about 30,000 years ago, humans had developed an impressive array of tools and weapons. Their hunting arsenal included stone axes with keen cutting edges and sharp spear points fashioned from bones and antlers. Groups living near rivers, lakes or the ocean had learned to make fish traps, as well as fish hooks and nets. In total, the human toolkit included more than 100 items. Humanity had also mastered fire; they now could summon it at will by generating sparks from flint and other stones or by friction. In addition, they soon learned to use it as a land management tool, setting fires regularly to make their hunting areas more accessible and to improve conditions for grazing animals.

A small, unheralded invention — the eyed needle, was as important as fire in allowing humans to populate the earth. The first needles probably were made by sharpening a sliver of bone or ivory and then boring a hole in one end with a small flint awl. With needles, humans could sew multi-layered, cold weather clothing. This simple implement, along with fire, enabled tropical humans to adapt to almost any climate on earth — including the frozen arctic. Humans probably invented the needle and learned to sew clothing about 75,000 years ago. This is about the same time that human head lice and body lice diverged genetically. Unlike humans, most animals support only one species of louse. This suggests that body lice arose to occupy the new environment created when early humans started to cover their bodies with clothing (Carr 2005).

In addition to an array of weapons, tools, clothing, and the ability to make fire, humans living 30,000 years ago had developed an impressive food technology. For example, they knew how to remove the toxic or bitter compounds that are in many nutritious, high energy foods. Examples are acorns and legumes in temperate regions and cassava in the tropics. A common technique was to place acorns, legumes, or other foods in a loosely woven bag and immerse them in a swiftly running stream for several days, allowing the flowing water to leach out the offending chemicals.

The ability to remove the bitter tannins from acorns was especially significant, since oak forests are so prevalent in parts of the Middle East, Europe, and North America. Oak forests became a major source of food for many hunter-gatherer groups outside the tropics. Interestingly, colonial records show that cooked acorns were one of the first foods offered by Native Americans to the Puritan settlers of New England (Philbrick 2006).

Cultural Evolution

True creativity and genius are bequeathed to only a few humans. But those of us not so blessed can be consoled; once our intellectual betters discover a new technology or new insight, most of us are able to understand it, apply it, and teach it to others. Doing so might take some education and hard work, but it is well within our capabilities. In the aggregate, the collective ability of humans to learn and teach might be nearly as important as individual genius. To understand the uniqueness of human learning, consider what happens each time you reach out and pick up something. When you do, certain neurons fire in your brain. If you then move the object from hand to hand, your brain will fire a different set of neurons.

Even more amazingly, if I watch you, the same neurons — mirror neurons, will fire in my brain. Not only can I watch you do something, I can "think" it along with you. Our brains are believed to be especially rich in mirror neurons compared with other species. The discovery of mirror neurons has excited many scientists, who believe they may be the key to human learning (Stamenov and Gallese 2002). Humans benefit from the genetic gift of being able to learn complex tasks and ideas by observing and listening to others; but for this to occur, they must be immersed in a culture where knowledge is retained and is available for learning.

Nearly everyone is familiar with biological evolution, which involves the transmission of genetic information from one generation to another. This information is coded on genes, which are simply chunks of genetic material — sequences of DNA. Our genes determine what we look like, how our body systems function, and to a considerable degree, how intelligent we are. When the gene pool undergoes a systematic change over time, evolution has said to have occurred. Genes that improve the survival and reproduction of a species tend to accumulate in a population over time. However, genes that are beneficial for one generation or for many generations can become lethal if the environment changes suddenly and drastically. That is why most of the species that once existed on earth now are extinct. Mother Nature is ruthless in weeding out the unfit.

This chapter has stressed the importance of individual intellectual and technological achievement. But another, perhaps even more important source of human power is our capacity to pool cognitive abilities and exchange knowledge, to create what British anthropologist Kenneth Oakley (1976) has referred to as the *social mind.* Writing a book is a very good example. A writer can benefit from the knowledge of people who lived centuries ago while at the same time gleaning information and insights from those still living. The transmission of knowledge and skills from one generation to another has been referred to as cultural evolution. Cultural evolution allows the store of human knowledge to become greater with each generation.

Cultural evolution can take place much faster than biological evolution. For example, humans learned to sew fur clothing much faster than they could evolve biologically to grow fur coats. Similarly, they learned to make fish hooks, fish traps and nets much faster than they could evolve to swim as fast or stay under water as long as fish. More importantly, because of the ability to transfer knowledge from mother to daughter and from father to son, each generation did not have to invent these technologies anew. By 15,000 years ago cultural evolution had enabled our species, still a tropical animal, to thrive in nearly every climate on earth.

This chapter has emphasized how our kind acquired an impressive array of physical and mental abilities as we made the transition from ape to human. The foot, the hand, a brain vastly more powerful than that of any other species, technology, language and culture — these endowed us with a formidable arsenal of coping and survival skills. We would expect early humans to have been, if not the masters of their world, at least to have been very successful. Scientists who began studying hunter-gatherer societies in the 1960s soon confirmed that this was so. These capable, self-reliant people were able to secure a good living even in the harshest environments, and they did so while working only a few hours each day. Anthropologists soon began using the phrase *Paleolithic rhythm*, which refers to the almost universal hunter-gatherer work schedule of "a day or two on, a day or two off."

This pattern has been well documented among the Bushmen tribes of Africa (Lee 1968). A woman can gather enough food in one day to feed her family for the next three days. She spends the remainder of her time resting, socializing, doing various tasks around camp, or caring for children. Males tend to hunt more frequently than women gather, but their day-to-day schedules are less predictable. A man might hunt every day for a week or more and then sit around camp doing nothing for the next two weeks. Such a work schedule is not unique to Bushmen. Pre-agricultural humans throughout much of the world followed the same pattern. Moderate effort enabled them to secure an abundant and nutritious diet. They were some of the best-fed people in history and perhaps the healthiest (Farb 1968).

The economics of hunter-gatherer life have been worked out quite thoroughly. Active hunter-gatherers work only about 20 hours per week on average and about one in three does no work at all. The approximately 30 percent of each band that is inactive consists of children and infants, the sick or injured, and those who have retired. It might seem odd to speak of retirement in a pre-agricultural setting, but in many hunter-gatherer societies, men quit hunting in their mid-fifties.

On average, about ten kilocalories are harvested for every kilocalorie spent in hunting or gathering. A 20 hour work week comes out to about three hours each day and three hours of hunting or gathering produce about 4,500 kilocalories. The average member of the band needs 2,500 kilocalories per day. This takes into account the fact that infants, children, and retired adults will consume fewer than 2,500 kilocalories, while some working adults will require more. Since one-third of the band does no work, each active member must meet his or her own food requirements (2,500 kilocalories) plus one-half the food requirements (1,250 kilocalories) of a nonworking member. Thus each active hunter or gatherer must produce a minimum of 3,750 kilocalories per day (2,500 plus 1,250). The typical production of about 4,500 kilocalories per day by each active worker is sufficient to meet this requirement with about a 20 percent margin of safety.

The numbers presented here are actually a worst case, or at least a bad case scenario. They are based on studies of the San or Bushmen living in the Kalahari Desert of southern Africa. It is hard to imagine humans living in a harsher, bleaker environment, yet the aboriginal people who made their homes there learned to live quite well with only moderate effort (Lee and Devore 1976). Procuring a living would have been even easier for people living in richer environments, such as the well-watered, fertile river valleys of western Europe, eastern North America, eastern China, or southeast Asia.

I have used the word band several times to refer to hunter-gatherer communities, which leads to a consideration of the so-called magic numbers issue. One of the magic numbers, 25, refers to the average size of a hunter-gatherer band. Although this number varied somewhat by location, the number of men, women and children in a typical band of hunter-gatherers was found to average about 25 throughout the world. There were reasons for this; a group of 20 to 40 individuals usually could field a hunting party of eight to 15 men. Studies of both chimpanzees and humans found that this is the most efficient size for a hunting group that lacks modern weapons (Ardrey 1976).

Interestingly, this number is still with us — consider the size of the typical military squad, the number of people on the Supreme Court, and the size of the typical city council. Although few could tell you why, most people would feel uneasy with a Supreme Court made up of 50 members. On the other hand, few would like a court consisting of only two justices. We simply feel more comfortable if the size of groups making important decisions for us is consistent with our hunting past. This is our Paleolithic conditioning at work.

When many people envision hunter-gatherers, they picture a forlorn group wandering the landscape at random, always lost and not sure if they will find food the next day. Nothing could be further from the truth.

Except for the few that followed migratory herds, all such groups had home bases from which they exploited a known territory. From a central camp, they could utilize the resources of more than 100 square miles simply by walking six miles in any direction. It is simple geometry: the area of a circle equals *pi* times the radius squared, or 3.14 times 6^2. This comes to approximately 113 square miles within a hunter-gatherer "working circle." Bands typically lived in the same territory for generations and knew the location of every nut tree, every fruit-producing bush or vine, every stream, and the habits of every edible animal native to the land or the water. Many bands would have two base camps — a summer and winter camp or a dry season and wet season camp. By moving about 12 miles in any direction, they had an entirely new range to exploit.

The other magic number, 500, refers to the tribal or linguistic group. A band of 20 to 40 people is a very efficient hunting and gathering unit, but is much too small for continued reproductive success. To ensure genetic survival, bands arranged themselves into larger, loosely knit tribal groups, sharing the same language and customs and having a common history, mythology, and religion. Although living apart, they would meet periodically for social interaction and to exchange an excess boy for an excess girl.

This provided greater genetic diversity, but early humans were unaware of this benefit. They simply knew that "marrying out" expanded their family ties and provided them with blood allies in other groups. It is believed that males usually stayed with their original band, while females migrated out — the same pattern followed by chimpanzees. This has advantages, since knowing that they are related increases cooperation among young males, a crucial factor in hunting groups.

This magic number also is still with us. In general, a town with a population below 500 or so often can get by without a formal police force. If it gets much larger than that, a town constable is needed to keep violence and conflict in check. Another example: as the number of students in a school exceeds about 500, discipline can become increasingly problematic.

Paleolithic Nutrition

Much of this chapter has been devoted to making the case for human competence, to counter a false perception that has led to major errors in thinking. Most of us do not find it remarkable that wild animals such as squirrels, deer, foxes, raccoons, and birds — almost any animal one could name, are well adapted to their lives in the wild and are easily able to feed themselves and their offspring. Wild animals rarely starve unless humans have disrupted their environment. Yet many people seem ready to believe that early humans, the most intelligent and resourceful of animals, were woefully helpless in the natural world, constantly teetering on the edge of starvation and extinction. This totally fictional feast and famine existence still is being used to "explain" the modern plague of metabolic diseases.

This chapter has provided a detailed chronicle of how biological and cultural evolution gave our species an unprecedented array of survival skills. This historical assessment is supported by eyewitness evidence that no existing hunter-gatherer group was ever seen to experience severe food shortage or starvation. On the other hand, no anthropologist or traveler has ever seen a fat hunter-gatherer. Instead, all members of every hunter-gatherer group encountered have been well-nourished but lean.

In most cases, fatness confers no advantage to an animal living in the wild. All land mammals, except for the few that hibernate, maintain a relatively constant lean weight. This is not an accident; in the natural world, survival is enhanced if as much of the body weight as possible is concentrated in strong bone and active fighting muscle. Early humans were no exception. The idea that they went through cycles of fattening up and then starving is fictional. Instead, they maintained a relatively constantly lean body weight because doing so enhanced their survival. Even lean, muscular humans can store enough fat to provide energy for more than a month. This would have been sufficient to cope with the kinds of short-term food shortages that early humans might have experienced. Early humans did not need to get fat, and doing so would have lessened their chances of survival, not improved them.

But that is a moot point. Considering the caloric density of the natural foods available to them and their habitual level of activity, pre-agricultural humans could not have gotten very fat even if they had wanted to. Harold Thomas, a Harvard physicist, demonstrated this for a group of African Bushmen (Pfeiffer 1969). He calculated the maximum kilocalories per person they could extract from their environment. He found that this could never exceed 3,200 kilocalories per person no matter how hard they worked. Most of the extra calories would be spent hunting, gathering and processing the extra food. Any weight increase would be minimal.

The same would apply to Australian aborigines, rainforest Pygmies, the early Shoshone of North America, or any other hunter-gatherer society. Their lifestyle and the nature of the food supply would not allow them to become fat, any more than a deer, antelope, or rabbit could become fat in the wild. Thomas discovered another interesting fact about the Bushmen. A group he studied had begun using metal-tipped arrows in the recent past, an innovation which had allowed them to improve their hunting success. But the extra food was not eaten and stored as fat by existing band members. Instead, their greater hunting efficiency allowed them to add four additional members to the group.

Although Paleolithic famines are not the cause of the obese phenotype, some of the modern tendency to become fat can be attributed to metabolic norms established during the age of hunting and gathering. One of these "metabolic norms" is the basal metabolic rate. This is the amount of energy required to keep the body running even if no activity is taking place. The brain, the liver, the digestive system, the lungs, muscles — all of the body's organs and systems require a certain amount of energy just to keep the life processes going.

The basic facts were laid out in a classic paper titled "A Biometric Study of Human Basal Metabolism" (Harris and Benedict 1918). The authors measured the basal metabolisms of 136 men, 103 women and 94 newborn babies. They found that men burned an average of 11.5 kilocalories for each pound of body weight just to stay alive. Women burned about 10.9 kilocalories for each pound of body weight. The 5.5 percent difference probably reflects the fact that men, on average, have a higher percentage of muscle mass relative to fat. A typical man weighing 150 pounds needs about 1,700 kilocalories each day to meet his basal metabolic needs; a typical woman weighing 120 pounds needs about 1,300.

In addition to the demands of basal metabolism, organisms expend energy moving around and doing work. The total energy expenditure of active adult humans typically ranges from 1.5 to 2.0 times their resting metabolic rate. Thus a moderately active man weighing 150 pounds will burn between 2,500 and 3,000 kilocalories each day. The same individual performing hard labor, such as intensive hunting in very cold weather, might burn 3,500 or more kilocalories per day. Recent measurements of men doing very hard work, such as cutting timber or mining coal, showed that their energy intake was sometimes more than twice their basal metabolic requirements.

Early hunter-gatherers were able to acquire their food while working only a few days each week, about the same energy output as a moderately active person today. On average, a hunting male would have taken in about 3,000 kilocalories of food energy per day and utilized perhaps 2,500 kilocalories, since the conversion efficiency is only slightly more than 90 percent for a high fiber diet. They could have acquired these calories by consuming more than five pounds of cooked wild game, since the average energy content of such food is about 600 kilocalories per pound. They would have needed to eat more than 15 pounds of raw vegetables or more than eight pounds of root crops such as potatoes, cassava, or yams in order to get the same number of calories. Nuts are a notable exception among wild foods in being very energy dense, yielding in excess of 2,500 kilocalories per pound. An early hunter-gatherer would have needed to eat less than one-and-a-half pounds of nuts to meet his daily energy needs.

The hunter-gatherer diet commonly included meat, vegetables, insects, shellfish, fruit, nuts, bird eggs, and any other tasty item they might encounter in the course of the day. It has been estimated that, overall, they consumed about four to five pounds of natural foods each day, depending on their body size and level of activity (Eaton et al 1988).

Most of the natural foods eaten by early humans are still around, and modern science allows us to measure their energy content with precision (Eaton and Konner 1985). The meat of wild animals is low in fat and energy density, generally averaging fewer than 700 kilocalories per pound (based on analyses of more than forty species). Wild plant foods are even lower in energy density, yielding fewer than 500 kilocalories per pound on average.

TABLE 4.1. KILOCALORIES PER POUND OF COMMON HUNTER-GATHERER FOODS

Food	Kilocalories per pound
Mushrooms	150
Starchy root vegetables (e.g., yams)	400
Fruit	200
Liver	650
Fish and shellfish	500
Brains	680
Pig	850
Small game (e.g., squirrel, rabbit)	750
Large game (e.g., deer, elk)	800
Nuts	3,000

Table 4.1 shows some representative foods known to have been eaten by hunter-gatherers and the amount of food energy provided by a pound of each. In most cases, the values in the table are averages of five or more foods. Data are from the U.S. Department of Agriculture food composition

database, which lists the energy and nutrient contents for thousands of foods. The website, called FoodData Central, can be accessed at fdc.nal.usda.gov.

On average, the mix of wild game, fish, shellfish, fruits, vegetables, and nuts eaten by hunter-gatherers yielded about 700 to 800 kilocalories per pound. Assume that a male hunter weighing 150 pounds needed to take in about 3,000 kilocalories each day. A simple calculation shows that he would have to eat about four to five pounds of food each daily (3,000 kilocalories divided by 700 kilocalories per pound). This would have equaled about two to three percent of his body weight. Our species lived as hunter-gatherers for more than 7,000 generations. During that time the average energy density of the human diet did not change appreciably. As humans moved into new regions, they encountered new plants and animals; but the meat of a deer in central Europe differed little in calorie content from that of a gazelle in the Middle East or a buffalo in North America.

This was the original set point at which humans stabilized and at which they remained lean but well-nourished throughout life. The data in Table 4.1 show just how hard it would have been for a hunter-gatherer to have become obese. It would be like the Vermont prisoners in Ethan Sims' overfeeding experiment. Can you imagine eating enough mushrooms, crayfish, animal brains, liver, yams, or wild pig every day to become obese? The most calorie dense food available to hunter-gatherers in any quantity was nuts, but it in order to become obese, one would have to eat well over a pound or more of them every day for an extended period of time.

Jean-Jacques Rousseau (1755) wrote that, in a state of nature, humans were strong, healthy and untroubled by the degenerative diseases of civilization.

> [M]en were strong of limb, fleet of foot and clear of eye... When we think of savages...and reflect that they are troubled with hardly any disorders save wounds and old age, we are tempted to believe that in following the history of civil society we shall be telling that of human sickness.

Rousseau's view of early humans is far too rhapsodic. They faced many dangers and hardships. Nevertheless, hunting and gathering provided a dependable, nutritious diet and a good life in general. Many anthropologists consider the time from about 30,000 years ago until the adoption of agriculture to have been the golden age of human nutrition. Nearly everyone could easily obtain sufficient calories and the natural foods they ate were high in protein, vitamins and minerals. Obesity was unknown and dental cavities were extremely rare. Far from the "solitary" and "brutish" existence portrayed by Thomas Hobbes, nearly all humans had enduring and comforting social networks, living all their lives in small bands consisting of relatives and companions they had known intimately from childhood.

Some have suggested that the biblical story of Adam and Eve reflects ancient memories of human life prior to agriculture and civilization. It is doubtful that there was a real Garden of Eden with two naked people and a wily serpent. It is more likely that this biblical metaphor speaks of an instinctual yearning for a way of life lost to us — for a time when humans lived free on the land, not yet hounded and enslaved by kings, priests and bureaucracy. As most readers will recall, Adam and Eve were exiled from the Garden of Eden after disobeying God and eating fruit from the tree of the knowledge of good and evil. For this act of disobedience, they were banished from paradise and condemned thereafter to earn their living from the soil by the sweat of their brow. In other words, they were exiled from Eden and sentenced to lives of hard labor — working in the fields.

On the surface, being sentenced to life as a farmer does not seem like so harsh a punishment, but it was. The next chapter chronicles the human suffering and decline in health that ensued after our species adopted agriculture. It is a tragic story — one marked by despotism, disease, hunger and malnutrition, all consequences of our exile. But an angry God did not exile us from paradise. Instead we exiled ourselves by forsaking the hunter-gatherer way of life and beginning to till the soil. Jared Diamond (1987) has referred to this fateful decision made about 10,000 years ago as the "worst mistake in the history of the human race." Whether agriculture can accurately be characterized as a mistake is open to debate; but it did have negative consequences for human nutrition and health, consequences with which we are still struggling today. Diamond's characterization raises an important question. If hunting and gathering was such a good life and we were so good at it, why did we give it up? Why make such a mistake?

Chapter 5. Peasants

Why, beginning about 10,000 years ago, did humans all over the world start to give up hunting and gathering to take up farming? Such a question would have seemed silly at one time because the answer seemed so self-evident. Until about 50 years ago, specialists from the many disciplines involved in prehistoric studies were of one mind, united in believing that agriculture had been an unqualified blessing. The benefits were so obvious, they agreed, that anyone given the opportunity would have embraced the farming way of life eagerly and as quickly as possible. There was little debate then about the "fact" that hunters and gatherers had lived short, miserable lives, engaged in a never-ending, often futile search for food. Tilling the soil had been a godsend, enabling early humans to settle down in one place and produce more abundant and more nutritious food with less effort.

These views remained virtually unchallenged until researchers began to study hunter-gatherer societies in detail and to reconstruct the history of early agriculture; then a different picture began to emerge. Farming did not lead to an easier, more abundant life. Studies from all over the world showed that early farmers often had to work as hard as or harder than hunter-gatherers did in the same kind of environment. The idea that cultivating crops is harder work than hunting and gathering might seem counterintuitive. But remember that farmers are responsible for all phases of food production; they must prepare the planting site, sow or plant the crop, weed it, protect it from marauding animals, and then harvest it. In contrast, hunters and gatherers had only to harvest the crop; they relied on a beneficent Nature to take care of the rest.

In addition to requiring more work, the shift from hunting and gathering to farming resulted in a more limited and less nourishing diet. Instead of ease and plenty, agriculture brought centuries of hard labor, along with disease and hunger for most of the people living on Earth. Until very recently, agriculture condemned the mass of humanity to consume those foods that were easy to grow or would produce a lot of calories, not necessarily those that tasted best or were the most nutritious.

Hunting and gathering has a downside as well; although it yields a more varied and nutritious diet, this lifestyle requires more land to support each person. This gives farming one major advantage as an economic strategy; it enables people to grow more food and support larger populations on less land. But most individuals derived few benefits from this; for them, agriculture meant more work, less freedom, less food security, a more monotonous diet, and poorer health.

Despite its human costs, the adoption of agriculture was a nearly universal event. A little more than 10,000 years ago everyone on Earth hunted and gathered wild foods. But by the time of Christ, nearly all of humanity had taken up farming. Why did hunter-gatherers abandon a subsistence strategy that had served them well for so long to take up a way of life that was less desirable in many ways? And why did so many people around the world — in areas as diverse as the Middle East, Mexico, the northern Andes, and East Asia — make the same "mistake" at about the same time? Referring to agriculture as a mistake, as I just did and as other authors have done, is both inaccurate and unfair. In terms one might hear in a courtroom, it assumes two facts not in evidence: first, that adopting agriculture was a deliberate decision, and second, that early humans actually had the luxury of choosing. Unfortunately, neither of these assumptions is true. Adopting agriculture was not a conscious, deliberate decision and there really was never a choice to be made (Cohen 1977).

About 75,000 years ago a huge volcano exploded on the Indonesian island of Sumatra, sending almost unimaginable amounts of dust and other debris into the atmosphere, darkening Earth's skies for years and plunging the world into a prolonged cold period. Some believe that this event nearly destroyed humanity, leaving only a few thousand survivors worldwide. Surprisingly, the near destruction and severe hardships resulting from this tragedy might have been beneficial in the long run. As the philosopher Nietzsche famously wrote, "*Was mich nicht umbringt macht mich stärker*," or "What does not kill me makes me stronger." That could have been the case here. The struggle to survive in the aftermath of this event might have forced the human race through the eye of an evolutionary needle, creating conditions in which only the most resourceful and most intelligent could survive, strengthening our species (Carr 2005).

Following this near-extinction, humans began to spread out and occupy new land, starting an era of population growth that continues to the present day. Survival of the species was no longer at issue. Instead, the problem became one of adapting to steadily increasing population growth. At first, the main response to greater numbers was out-migration, with local groups spinning off daughter colonies to settle adjacent lands. This is similar to what happened in colonial America; as more babies were born and more immigrants arrived from Europe, pioneers continuously moved west to fill up new territory. Early humans followed the same strategy. First they expanded out of Africa, occupying the Middle East. Then, as populations continued to grow, they colonized more and more of the world.

By 15,000 years ago our species had spread to every continent that would support the hunter-gatherer lifestyle, with an estimated population of five to ten million. As the world began to fill up, colonizing new territory became less an option, so other ways of feeding additional people had to be found. A number of strategies came into play, including the improvement of hunting and gathering technology and discovering processes to turn once-inedible foods into new sources of calories and protein. The mastery of fire, which led to cooking, was discussed in the previous chapter. Many calorie-rich plant foods such as potatoes, cassava, grains and legumes cannot be eaten in quantity when raw. But cooking makes them palatable and easy to digest.

The mastery of fire made a number of high energy foods available for the first time, in effect increasing the food supply and enabling the same area of land to support more people. We also have pointed out the importance of learning to remove the toxic or bitter compounds from certain high energy foods such as acorns in temperate regions and cassava in the tropics. Along with cooking, these technologies greatly increased the effective food supply. Other important breakthroughs included new hunting technologies, including spears, bows and arrows, snares, nets, fish poisons, and fish hooks. In addition to new technologies, humans soon developed more clever and elaborate hunting strategies, including the use of fire to drive game.

Increasing populations forced early people to broaden their food preferences. Foods that once had been ignored or rarely eaten became a regular part of the diet. Consumption of certain categories of foods increased substantially, notably those that were plentiful, high in calories, and easy to gather — even if they did not taste all that good. Root crops such as cassava, taro, and potatoes; and grains such as wheat, corn, and rice are examples. Humans did not choose these foods because they were particularly delicious. Necessity alone made them an important part of the diet; they were not eaten in quantity until there was a shortage of more preferred foods such as meat, nuts and fruits.

Hunters and gatherers also began to encourage "gardens" of certain preferred food plants near their camps. They had an almost encyclopedic knowledge of the vegetation growing in their territory, so the way in which plants reproduce was no mystery to them. They knew the conditions under which various plants grew best and when the fruits, seeds, or nuts were ready to eat. They learned to purposely discard or bury the seeds of favored foods near their camps or, in the case of crops like bananas, cassava, or fruit trees, to bring cuttings or small seedlings back to camp and put them in the ground. Some crop plants probably invaded human settlements on their own, taking advantage of disturbed sites such as pathways, burned areas, refuse heaps, or abandoned villages.

Early humans soon learned to improve their hunting grounds as well. It became a common practice to burn grasslands or the forest undergrowth on a regular basis. The Indians of North America used fire to reshape entire ecosystems, replacing forest with savanna in much of the eastern United States. Early settlers in Ohio were able to drive their carriages and wagons through many areas of open woods where annual fires set by the Indians had created vast parks for buffalo, deer, and other preferred species. The early Dutch inhabitants of New Amsterdam (later New York) enjoyed a holiday each fall traveling up the Hudson River to enjoy the fiery spectacle as the Indians conducted their annual burning (Mann 2002).

Similar practices were common throughout the world. Periodic burning enhanced grazing, improved hunting access, and probably led herd animals to begin concentrating in the fire-maintained areas near human camps. Groups of humans probably fed off the same herds year after year, learning to recognize individual animals. As herds began to cluster next to settlements, they became more subject to predation by humans, but were protected from other predators, such as lions or other large cats, marauding packs of wolves, hyenas, or wild dogs. The herd animals learned to recognize humans as a constant part of their world. This close association was a precursor to eventual domestication.

Farming was not the result of a sudden decision or a single conceptual breakthrough. Instead, it was the gradual accumulation of new habits, resulting from many individual adjustments taking place over long intervals of time as populations increased. Mark Cohen published a book in

1977 titled *The Food Crisis in Prehistory: Overpopulation and the Origins of Agriculture*. Most of the discussion presented here on the gradual transition from hunting and gathering to agriculture is based on this very comprehensive work. In most early societies, hunting and gathering coexisted with limited agriculture for hundreds, sometimes even thousands of years. Succeeding generations were unaware that, slowly but surely, they were abandoning their hunter-gatherer existence and becoming farmers. Two well-documented examples of this come from the Middle East — one dealing with the cultivation of grains and the other with the domestication of animals.

Archaeologists studied the agricultural history of an early village in what is now southwestern Iran by examining old refuse dumps layer by layer. About 9,000 years ago cultivated plants accounted for only five percent of the seeds recovered. The remaining 95 percent were seeds that had been gathered from wild stands. Within 1,000 years the percentage of cultivated seeds had increased significantly, but still represented only 40 percent of the total (Budiansky 1999). This illustrates the normal course of events; early societies adopted agriculture not in a transforming flash, but very slowly over many hundreds of years.

Excavation of an ancient settlement in what is now Syria showed that animal domestication followed a similar pattern of long-term incremental change. The bones of domesticated sheep and goats first appeared about 9,500 years ago, when they accounted for about ten percent of the bones found at the site. The bones of wild gazelles made up most of the remaining 90 percent. Gazelles were once numerous in the Middle East and the large migrating herds were a major source of food for early inhabitants (Legge and Rowley-Conwy 1987). As in the case of plants, animal domestication increased at a very slow rate. It took about 1,000 years for the bones of domestic sheep and goats to make up more than half of the total. These examples show that agriculture did not suddenly replace hunting and gathering in the Middle East; instead the transition was slow and incremental, with agriculture playing only a supplemental role for many centuries (Redding 1988). It is likely that the transition to agriculture followed a similar pattern in the other major world centers, such as China, Mexico and Central and South America.

In nearly all cases, people adopted agriculture only out of necessity, holding on to the hunter-gatherer life as long as possible. Among groups who practiced mixed economies, there seemed to be a distinct preference for hunting and gathering over farming (Murdock 1968). This should not be surprising. Agriculture typically requires hard, concentrated labor that might not be rewarded until months later. Hunting and gathering have a more immediate payoff and, in addition, are more leisurely, interesting and entertaining pursuits. When given the chance, agricultural people often would go back to hunting and gathering.

In colonial America, there was a continuing problem of men leaving European settlements to live with the Native Americans. Only a few years after Virginia was settled, more than 40 male and a few female colonists had left to go native. Officials became so alarmed that they instituted severe penalties for *Indianization*. This term came into use during the 1600s, a time when New England's Cotton Mather asked with concern, "How much do our people Indianize?" A French observer wrote in 1782, "there must be in the Indians' social bond something singularly captivating, and far superior to be boasted of among us; for thousands of Europeans are Indians, and we have no examples of even one of these Aborigines having from choice become European" (Farb 1968).

Agriculture and Disease

People everywhere resisted giving up the hunting and gathering life for as long as possible. In retrospect, they were wise in doing so; although agriculture made greater population growth possible, it exacted a huge price. Its legacy included disease, chronic malnutrition, and centuries of exploitation and oppression. The story of Pandora's Box could well be a metaphor for the human tragedies unwittingly unleashed by agriculture, as could John's apocalyptic description from Revelation 6.

> And I looked, and behold a pale horse: and his name that sat upon him was Death, and Hell followed with him. And power was given unto them over the fourth part of the earth, to kill with sword, and with hunger, and with death, and *with the beasts of the earth.*

Shakespeare wrote in Antony and Cleopatra that "Nature teaches beasts to know their friends." The great bard's observation is insightful in the sense that certain animals are able to recognize a kindred spirit in humans. Nearly all of the animals that humans domesticated lived originally in herds, packs, or other social groups much as early humans did and were similar to humans in having a dominance hierarchy. Dominance hierarchies typically are established by fighting, but their ultimate purpose is to establish an ordered system and minimize conflict. Once dominance is established, a subordinate animal will give way to a superior. Many animals can be taught to recognize humans as dominant and to enjoy being groomed or petted by them. Considering the social conditioning of herd animals, it is no surprise that a man can easily control a horse, a cow, or even an elephant many times his size (Budiansky 1999).

Within a period of only a few thousand years, social animals such as horses, cows, goats, sheep, pigs, chickens, and dogs all moved in with humans. Unfortunately, they brought their diseases with them; nearly every infectious disease that affects humans had its origin in animals. For example, the flu virus had its ancestral home in birds and swine. As soon as they were domesticated, pigs and birds began exchanging constantly mutating flu viruses with humans. Periodically a variant arose for which our species had little or no immunity, resulting in an epidemic or pandemic, such as the killer flu of 1918. Measles in humans is related to a virus that causes distemper in dogs. A virus similar to smallpox causes vaccinia in cows and pox infections in birds and pigs. Bubonic plague was spread by flea-infested rats as they flocked to early human farms and settlements, taking advantage of the shelter and grain available there (Karlen 1995, McNeil 1998).

Tuberculosis is an especially interesting case. Bovine tuberculosis is an ancient disease of cattle caused by an organism named *Mycobacterium bovis*. When people domesticated cattle and began to drink their milk, some of them acquired this microbe, which learned to thrive in the human bowel, in the lymph glands of the neck, and in the spine. The result was Pott's disease in humans, which is a pretty nasty condition, but things would get even worse. About 4,000 years ago, *Mycobacterium bovis* mutated into a form that eagerly took up residence in the moist, oxygen-rich tissues of the human lung. From the lungs it was easily spread by coughing or close contact. Tuberculosis was on its way to becoming one of man's greatest scourges. As late as 1840, this one disease was the single largest cause of recorded deaths in English cities. Tuberculosis remained a major killer throughout the world until nearly the middle of the 20th century, when it was largely eliminated in the developed world. But this old enemy now appears to be making a comeback.

Influenza, smallpox, plague, and tuberculosis are but a few of the diseases that followed the domestication of animals. Biologists have compiled lists of diseases the human race acquired from various domesticated species. The total includes more than 40 diseases each from cattle and pigs, more than 30 from horses, more than 25 from poultry, and more than 60 from dogs (McNeill 1998). Many of these diseases appeared early in human history. A priest in a Hittite kingdom lamented around 1400 BC,

> ...men have been dying in my father's days, in my brother's days, and in mine since I have become a priest of the gods...The agony in my heart and the anguish in my soul I cannot endure any more (Pritchard 1969).

Deadly animal-borne diseases so devastated early agricultural societies that they were assumed to be punishment meted out by an angry god. Why else would a people be so afflicted? Consider the following promise made to the Israelites in Deuteronomy 7:15.

> The Lord will take away all sickness from you; he will not bring upon you any of the foul diseases of Egypt which you know so well, but will bring them upon all your enemies.

Despite poor nutrition and disease, populations increased under agriculture, mainly because of closer birth spacing. Even an undernourished and disease-ridden population can grow rapidly if women begin reproducing early and have children at frequent intervals. Agricultural mothers did not breastfeed their babies for as many years as hunter-gatherers; instead they weaned their babies early on a gruel or porridge made from grain or soft, starchy vegetables. Earlier cessation of breastfeeding enabled them to become pregnant more often and to have more children during their lifetimes than women in hunter-gatherer societies.

The result, in most cases, was explosive population growth. Populations in a given area typically increased ten or 20 times within only a few hundred years of adopting agriculture. The early farming population of the Middle East is believed to have increased from fewer than 100,000 to more than three million in only 160 generations. Other centers of agriculture throughout the world experienced similar surges in population (Farb 1978).

In the spirit of "Nature teaches beasts to know their friends," there are some interesting parallels between man and his best friend, the dog. It is widely assumed that dogs were deliberately domesticated by humans for use in hunting, but this is untrue. Actually, dogs domesticated themselves. They evolved from wolf-like ancestors after moving into areas surrounding the refuse dumps of human villages. Scrounging for waste discarded by humans was a much easier life than chasing down and killing their food. Adapting to this new niche caused many physical changes. Compared with wolves, village dogs the world over have smaller bodies, less massive jaws, and smaller teeth. Even after accounting for less body mass, they have smaller brains than their wolf ancestors and are not as intelligent. Reproduction was altered as well; whereas wolves have only one litter of pups a year, village dogs often have two (Coppinger and Coppinger 2001).

Domestication had similar effects on people. Compared with the hunter-gatherers who preceded them, farmers were smaller in stature, had smaller jaws and teeth, and gave birth more frequently, about every two years instead of four. Finally, although culture and technology conceal this embarrassing fact, we modern humans have smaller brains than our more recent hunter-gatherer forebears and might even be less intelligent.

As agricultural populations continued to grow, economies and cultures specific to certain food crops developed in various parts of the world. Middle Eastern agriculture was based mostly on wheat, while corn was the main crop in Mexico and Central America. Chinese agriculture began with millet production on the deep, easily-cultivated loess soils of the northeast. This was followed a few thousand years later by intensive, irrigated rice farming. Potatoes, legumes and other foods were important in some regions, but three crops — corn, rice and wheat, were to become dominant, eventually providing most of the calories consumed by the human race. These three crops varied greatly in their evolutionary and cultural histories, and had different nutritional effects on the human populations that consumed them.

Wheat

Although wheat is native to the Middle East, a cataclysmic event in distant North America is suspected of triggering its domestication. About 12,000 years ago a melting ice sheet collapsed in Canada, allowing the waters of giant glacial Lake Agassiz to begin pouring out. Vast amounts of water drained out of the lake through the Saint Lawrence Seaway, cooling the waters of the Atlantic and the air above it. As a result, the climate of the Middle East became both colder and drier. Up to that time, people living there had enjoyed an abundant and diverse diet of small game, fish, shellfish, pistachios, acorns, and gazelles, supplemented by wild wheat and barley in some areas. But a colder, drier climate caused many of their food sources, including oak forests and gazelle herds, to become less abundant; increasingly, the inhabitants had to rely on wild grains for survival. At first they relied totally on gathered seeds, but population pressure soon forced them to start harvesting wild grain and planting it.

Modern wheat appeared in the Middle East about 10,000 years ago from mergers among several wild grass species; in only a few thousand years, this new plant developed three properties that made it a desirable food crop. First, the seeds became unusually large, making the grain both more palatable and easier to process. Second, the rachis, to which the seeds are attached, mutated to a form that held the seeds more tightly; as a result, whole ears of grain could be harvested, instead of individual seeds scattering on the ground. Third, the leaf-like structures that cover each seed became looser, so they could be more easily removed to free the naked, edible grain. Wheat became the staple food in the Middle East and later in Europe, eventually spreading to European outposts such as Canada, Argentina, Australia, and the United States.

Corn

When Spanish conquistadors first ventured into the western hemisphere nearly five centuries ago, they "discovered" a world that was more populous than Europe at the time and had cities much larger than London, Paris, or Madrid. The calories that fed this large population came mostly from corn, which was the "staff of life" in the Americas, playing a role similar to that of wheat in the Middle East and of rice in Asia. The peculiar grass we call corn owes its existence to a series of lucky events. Evolving from an ancestral plant called *teosinte*, corn is believed to have sprung from a founding population of fewer than 30 individuals through a series of mutations that took place within ten or fewer generations. These mutations occurred at only one place, in the Balsas River valley in southern Mexico more than 5,000 years ago. Fortunately, humans found this early variety of corn and began to cultivate it.

Corn can produce such a large volume of food energy per unit of land because it is able to carry out photosynthesis with such great efficiency. It is one of a select group of plants that botanists call C-4, as opposed to most other plants on Earth, which are C-3. This designation relates to the way the plants carry out photosynthesis, the process by which plants use the sun's energy to convert carbon dioxide and water into sugar. To begin the process, plants first must take carbon dioxide out of the air, and this comes at a biological cost. Plant leaves have numerous microscopic orifices called stomata through which carbon dioxide is taken in. But every time a stoma opens, the leaf loses water to the atmosphere. This can be a problem, especially in hot, dry areas.

Corn and a few other plants have partly overcome this dilemma by evolving the ability to "fix" more carbon each time they open their stomata. By limiting water loss during carbon recruitment, C-4 plants are able to produce more biomass, including edible grain, with a given amount of rainfall. The designation C-4 refers to the fact that, during photosynthesis, such plants initially create sugars that have four carbon atoms, as opposed to C-3 plants, which produce compounds with only three carbon atoms (Pollan 2006).

This natural advantage of corn, along with clever management strategies developed by early farmers, made it possible for this crop to support large populations. One such trick was the practice of planting corn and beans together. The tall upright corn plants provided physical support for the climbing bean vines, enabling them to get sufficient sunshine. In return, the leguminous beans increased corn yields and permitted continuous cropping by adding nitrogen to the soil. The agronomic partnership of corn and beans also benefited human nutrition, since the two foods eaten together provide all of the amino acids humans require to make protein.

Europeans quickly saw the advantages of corn and adopted it enthusiastically. The numbers were compelling. Each corn seed planted typically returned more than 150 kernels and often as many as 300 in the resulting crop. The payoff from wheat was not nearly as good; each wheat seed sown returned less than 50 at harvest time. It is no surprise that corn became the crop of choice in early America. Even today, corn produces 50 percent more grain per given area than rice and exceeds the productivity of wheat by an even greater margin.

Rice

In the minds of most people, the words rice and China naturally go together. No one knows for sure when they first became rice farmers, but the Chinese had made significant progress in taming the immense Yellow River floodplain by about 600 BC. They drained swamps and marshes and constructed a vast system of dikes and canals, eventually creating an almost unbroken expanse of rice paddies along this great river. Rice farming soon spread throughout much of China and into other parts of Southeast Asia. Like corn, rice has some special advantages as a food crop. Since it can tolerate wet conditions, deliberate flooding during certain phases of the growth cycle can be used to control competing weeds and to provide additional water to enhance growth and grain production.

Rice also benefits from the presence of an aquatic fern that floats on the water surface of flooded areas in many parts of the world. This plant, of the genus *Azolla*, is commonly called mosquito fern. It has tiny leaves and actually looks more like moss than the ferns one commonly sees in the woods. The leaves of the mosquito fern are home to numerous colonies of a certain bacterium that is able to take nitrogen from the air. Mosquito fern and its associated bacteria often form solid masses on the surfaces of rice paddies.

Azolla grows very fast, reportedly doubling its mass in as little as three days under good conditions. When the fern dies, it provides large amounts of nitrogen to the growing rice plants, especially if the rice paddy has dried out. This useful plant enabled Asian rice farmers to produce rice on the same fields year after year for centuries.

Although not as productive as corn, rice can yield much more edible grain and support larger populations on a given area than wheat. But the complexity and labor-intensive nature of rice culture requires a large number of workers. In the absence of animal power or machines, growing paddy rice required at least five times more human labor per unit area than upland wheat or corn. The complex water management necessary for rice production also required a much greater degree of cooperation, planning and management. It is not surprising that rice-growing regions of the world quickly developed very large peasant populations living in strictly controlled societies (Wolf 1966).

Around the year AD 1000, the rice-eating population of Japan was four times larger than the wheat-eating population of England. By 1800, the population of China, at more than 300 million, was twice that of Europe (McNeil 1998). Even today, rice feeds more of the world's people than any other crop. It is unique among the major food crops in that most of it still is grown for local consumption, with less than five percent reaching the world market. Largely because of rice, modern China manages to feed one out of five people on the planet using less than seven percent of the world's land (Williams 1996).

The Enduring Peasants

Farming soon brought with it oppressive economic systems in which the mass of humanity was deliberately deprived of food by those who had assumed power over them. Persistent, hard labor at starvation wages became the lot of most humans only a few thousand years after the adoption of agriculture. Nearly everyone now living in Europe and America is of peasant stock. The same is true of modern Chinese, Japanese, Mexicans, Indians, Southeast Asians and South Americans — of all of the world's people except for a few small groups of subsistence farmers and hunter-gatherers that survived into modern times.

For generations our ancestors were at the mercy of whoever owned the land on which they lived. In many cases a ruler had bequeathed it to some favored subject in the past as payment for political support or military service and the land had since been handed down from generation to generation. Peasants typically came with the land and often were forbidden by law from leaving it. The products of their labor beyond the most meager subsistence were taken as "rent" or "taxes" by a political ruler, a wealthy landlord, the nobility, or the church.

Peasant economies, organized systems based on permanent human bondage, began about 5,000 years ago in the Middle East and China and about 3,000 years ago in parts of Mexico and Central America. It soon became nearly universal. A distinction must be made between peasants and subsistence farmers. Subsistence farmers, who still exist in some isolated parts of the world, typically work only a few hours each day to meet their caloric needs and feel no compulsion to work longer. The rest of their time is spent in leisure. In contrast, the peasant is robbed of his leisure — compelled to work long, hard hours in order to pay tribute or "rent" to an outsider who has usurped power over him. Peasantry requires a formal social order in which some men can use power to demand payments from others, based upon one pretext or another. Peasants are forced to produce a surplus, which then is confiscated by the dominant group to underwrite a higher standard of living for itself.

In addition to being calorie-deficient, the most striking thing about the peasant diet is that it contained almost no animal protein. A proclamation by King Henry IV of France upon ascending the throne in 1589 is particularly telling: "If God grants me the usual length of life, I hope to make France so prosperous that every peasant will have a chicken in his pot on Sunday." King Henry reigned more than 20 years and embroiled the country in a succession of expensive wars, while at the same time maintaining a lavish lifestyle at his court. But somehow he never was able to improve the diet of peasants (Farb 1978). Historically peasants the world over have suffered from chronic protein deficiency. In addition to lack of protein, generation after generation grew up on diets deficient in total calories, essential fats, vitamins and minerals.

Regardless of where peasants lived, their lot was pretty much the same — a life of permanent bondage, taxation, forced labor, military conscription, and usurious rent. Peasants were systematically starved because the products of their labor beyond the most meager subsistence were taken as "rent" or "taxes" by a political ruler, a wealthy landlord, the nobility, or the church. A British historian described the plight of Chinese peasants in the 1930s (Ward 1976).

> [T]he rural population suffers horribly...It is taxed by one ruffian who calls himself a general, by another, by a third and when he has bought them off, still owes taxes to the government...There are districts in

> which the position of the rural population is that of a man standing permanently up to the neck in water, so that even a ripple is sufficient to drown him.

One might wonder why peasants would put up with such treatment. They simply had no choice. A brutal system of "laws," armies of thugs, and — in some societies, religious bureaucracies keep them in line; woe unto any peasant so reckless as to "steal" from the lord or refuse to pay his share of "taxes." In Europe, the early Catholic Church played a role by assuring devout peasants that the status quo, however unfair or inhuman, had been sanctioned by God. In the allegorical poem *Piers Plowman*, Piers feels obligated to "sweat and sow for us both" and, in return, he expects the local ruler to "keep holy church and myself from wasters and wicked men." A similar view is expressed by the parson in Chaucer's *Canterbury Tales*, written in the late 1300s. According to the parson, God had "ordained that some folk should be higher in estate and degree and some more low, and that everyone should be served in his estate and his degree."

It is distressing to think that early church leaders participated in and profited from such an inhumane system; but religious bureaucrats of the time did not hesitate to warn peasants in sonorous terms that, because of their lowly birth, God required that they accept their condition, serve their masters, and hope for salvation in the next world. Peasants were ordered to work hard as part of their Christian duty despite being brutalized and half-starved.

Systems that forced hungry peasants to live in squalid conditions for generations were further justified on the grounds that peasants were lesser human beings and deserved their fates. Even today, the word peasant conjures up, in the minds of many people, the vision of a crude individual with limited intelligence. I checked a number of dictionaries and jotted down some of the more commonly used synonyms for peasant. Here are some examples: churl, clown, bumpkin, hobnail, hick, clodhopper, yokel, hayseed, rube, drone, boor, and commoner.

Long after they were no longer peasants, poor people all over the world, including Europe and America, continued to eat mostly peasant foods. As late as 1900, agricultural production barely met human needs, so nearly all of the world's grain was consumed directly by humans, with only about ten percent fed to cattle, hogs, or other meat producing animals. In 1850, working class families in England got 70 percent of their total calories from bread and potatoes alone. A survey conducted in 1892 found that more than 80 percent of the children in one English community had eaten only bread for 17 of their last 21 meals. As late as 1910, working class urban dwellers in both the United States and England still were getting nearly half their calories from bread and potatoes. At the outbreak of World War II, the poor of England still were eating mostly bread, with each person consuming about a pound and a half daily (Mount 1975). Working class and poor people in other parts of the world, such as Asia, Africa, and South America, also ate mostly peasant foods well into the 20th century, with rice or corn making up the bulk of their diets.

Table 5.1 lists some representative foods that have been eaten by peasants for thousands of years and the amount of food energy provided by a pound of each. Data are from the U.S. Department of Agriculture food composition database, which lists the energy and nutrient contents for thousands of foods. The website, FoodData Central, can be accessed at fdc.nal.usda.gov.

Table 5.1. Kilocalories per pound of common peasant foods

Food	Kilocalories in one pound
Rice (whole grain)	500
Noodles/pasta	590
Corn (boiled or roasted)	385
Corn grits	275
Corn tortillas	510
Beans/lentils	540
Wheat bread	545
Wheat cereal	280
Oatmeal	250
Potatoes	395

The data in Table 5.1 indicate that peasant diets probably provided about 450 to 550 kilocalories per pound. The main reason for the relatively low energy density of the peasant diet was its low-fat content. It has been estimated that, for peasant diets worldwide, fat rarely accounted for more than ten percent of calories. In contrast, fat is believed to have accounted for 20 percent or more of calories in hunter-gatherer diets. As a result, the overall energy density of the peasant diet was much lower than that of hunter-gatherers.

The peasant diet also contained much less protein, vitamins, and essential fats, so it is no surprise that only a few generations after taking up farming, much of the human race became much shorter. Most hunter-gatherers living 30,000 years ago were as tall as modern humans, about 5 feet, 9 inches for men and 5 feet, 4 inches for women. In contrast, the typical male peasant, especially in corn and rice growing cultures, was only a little over five feet tall. According to historical records, the average height of the feared Japanese Samurai was 5 feet, 2 inches. This does not mean, of course, that they were not fearsome, formidable warriors.

Hunger and Revolution

In the fall of 1793 at the age of 38, Marie Antoinette was hauled to her death in a crude two-wheeled cart while crowds lining the road jeered and mocked. A shocking sight, the once proud and beautiful woman was now pallid and gray, her body filthy and her hair roughly shorn. Even the spiteful crowd must have been shocked by the haggard figure passing before them. Months of abuse in a dark, dirty prison cell had aged her far beyond her years. Arriving at the appointed place, Marie Antoinette was dragged roughly from the cart and her head was cut off to the accompaniment of resounding cheers.

But most readers will have little sympathy for her. After all, we know the story; Marie Antoinette was perhaps history's best example of cruel, self-indulgent royalty, living in luxury while the common people starved. When told that the populace was desperate because there was no bread, she replied disdainfully, "Let them eat cake"— a harsh statement that history has neither forgotten nor forgiven. The problem is, it most likely never happened; the whole thing probably was propaganda circulated by her enemies to inflame public opinion against her. They most likely appropriated this now infamous remark from Jean-Jacques Rousseau's *Confessions*, "At length I recollected the thoughtless saying of a great prince who, on being informed that the country had no bread, replied, 'Let them eat cake'."

On the day Rousseau penned these words, Marie was still a carefree child approaching 11 years of age and living happily in Austria. Hopefully, she had no premonition of the violent and degrading death awaiting her in France or that a philosopher of that country was writing words that would help seal her fate. She was neither the first nor the last historical figure to be deliberately and maliciously misquoted, but few have suffered for it as she did.

Marie Antoinette might not have been the monster we have been led to believe. There is evidence that she was concerned about the suffering of the French people and was, in her own way, trying to help. For example, she wore potato flowers in her hair as a way of encouraging distrustful peasants to grow this strange new crop that many of them feared. Growing potatoes could have reduced the nation's hunger. Just one acre would have fed a peasant family of five and a pig as well. Potatoes grow well in a wide range of soils and are easy to cultivate, requiring only a hoe or a spade. They mature quickly, producing an edible crop in less than four months, compared to six months or so for grain crops.

But most French peasants stubbornly refused to grow them, believing they had good reasons. After all, potatoes came from heathen lands across the sea and, unlike wheat, were not mentioned anywhere in the Bible. Surely the Lord did not intend for humans to eat them. Their manner of growth also was suspect. They reproduced underground, in the dank, fetid darkness, in furtive ways that surely were abominable in the eyes of both man and God.

If French peasants had adopted potatoes earlier, much hunger might have been avoided, the French revolution might never have occurred, and Marie Antoinette might have lived to a happy old age. But the peasants failed to embrace potatoes and Marie did not survive her fourth decade. Instead, she suffered and died an early death, not necessarily for committing heinous acts, but as the hapless victim of a reality that no ruler should ever forget: The first duty of government is to make sure that the populace is fed. To ignore this is to court disaster, since the ruler of every country, regardless of the form of government, is just nine missed meals away from violent overthrow. Uneasy lies the head that wears the crown — or the mantle of presidential power. Charles Dickens, in *A Tale of Two Cities*, gives a human face to the hunger and desperation that culminated in scores of bloody heads tumbling into a waiting basket below the busy guillotine.

> [T]he children had ancient faces and grave voices; ...Hunger stared down from the smokeless chimneys, and started up from the filthy streets... Hunger was the inscription on the baker's shelves, written in every small loaf of his scanty stock of bad bread; at the sausage-shop, in every dead-dog preparation that was offered for sale...Hunger rattled its dry bones...

Food shortages and famine were not unique to France. Most of Europe suffered from malnutrition and hunger throughout the 17th and 18th centuries. But other countries were fortunate in escaping the bloodshed and turmoil that convulsed France in the late 1700s. Although food shortages remained a constant threat to Europe, increases in wheat production, along with the wider adoption of corn and potatoes as food crops, helped to sustain an ill-nourished but growing population. Importation of nitrogen fertilizer from overseas in the form of guano and nitrate salts also helped. But things were looking very bleak as the 20th century neared.

In an 1898 address to the British Association for the Advancement of Science, the eminent chemist Sir William Crookes warned that worldwide famine loomed as early as 1930. He offered this stark assessment: "England and all civilized nations stand in deadly peril of not having enough to eat... [W]e are drawing on the earth's capital, and our drafts will not perpetually be honored." Crookes sounded this alarm because rapidly growing populations were already using nearly all available land and were exhausting their sources of fertilizer. The conclusion was unavoidable: they would soon outgrow their food supply. It was clear something needed to be done quickly or famine and social unrest were imminent (Crookes 1899).

Fortunately for us, Crookes' fears did not come to pass. Instead of struggling with starvation and civil unrest in the 20th century, the industrial economies increased their food production almost exponentially. A few years after World War II, the United States and much of Europe struggled, not with shortages, but with a glut of agricultural production. They found themselves in the enviable position of trying to dispose of excess grain, milk, cheese, butter and, in some cases, even meat. This situation came about despite the fastest rates of population growth in human history. In addition to feeding many more people, the industrial economies greatly increased the amount of food available to each person. Much of the industrial world now faces a dilemma that is just the opposite of what Crookes feared. We live in societies where the potential for food production, at least in the short term, is practically unlimited. Instead of want and starvation, more than half of humanity is on its way becoming obese and diabetic. Despite an almost unbelievable increase in population, much of humanity now suffers from the over consumption of fat, sugar, and calories. The modern industrial diet is causing widespread and possibly irreversible declines in human health. This is ironic, since advances in agricultural production and in food technology should have made us happier and healthier.

Nowhere is this more striking than in the United States, where some experts predict that the rate of overweight or obesity could exceed 90 percent by 2050. We soon could become a nation in which obesity and diabetes, along with heart disease, blindness, kidney failure, and amputations are considered a normal part of life. The abundance of the past century seems to have cursed us, but why and how did it happen? This question cannot be answered by focusing on the present or just on the last century. In order to understand the situation now confronting us, we need to start at the beginning, at the very founding of our nation. Accordingly, the next chapter describes the arrival of the Pilgrims at Plymouth Rock in 1620 and how they survived in the New World.

Chapter 6. An American Tale

For many Americans, 1620 marks the mythical beginning of our country. That is the year the Pilgrims arrived at Plymouth Rock in New England. Although others had come here much earlier, the Pilgrims were the first group to come ashore with the clear intention of staying. They arrived in America committed to becoming farmers, fishermen, or tradesmen, and making a permanent home on this new continent. A second date, 1920, marks America's transition from a rural, largely agricultural nation to one that is industrialized and urban. The 1920 census reported for the first time that our nation was home to more than 100 million people. Equally significant, 1920 was the first year in which a majority of Americans lived in cities or towns.

If the Pilgrims' arrival in 1620 represents the beginning of European settlement in America, 1920, three centuries later, marks its end. The closing of the American frontier had been proclaimed years earlier, right after the 1890 census, but this announcement was premature. New farmers continued to settle land in the Great Plains and West during the first two decades of the 20th century. The true end came shortly before 1920. By that time, America's farm population had peaked and the supply of public land open for homesteading was truly exhausted.

In addition to being a demographic milestone, 1920 is important for another reason; it marks the beginning of a scientific and technological revolution that transformed American life. After that year, industrial and scientific methods once confined to industry were increasingly applied to the growing, processing, distribution and marketing of food. The age of industrial or "factory" food was born. Increasing rates of obesity and diabetes, along with the debilitating conditions that accompany them, such as heart disease, strokes, blindness, kidney failure, and nerve damage leading to amputation are some of the regrettable and unforeseen consequences.

Sus scrofa

Although rarely, if ever, mentioned in history books, the lowly pig, or *Sus scrofa*, was a keystone species in the American diet for 300 years. Any accounting of food and nutrition during the early years of our nation requires that the pig be given its due. Pigs have a long history in America, having come here well before the Pilgrims. Spanish explorers brought pigs with them on all of their journeys. Christopher Columbus introduced pigs to the Caribbean islands on his second voyage in 1493 and Pizarro took a herd with him to Peru in 1531 (Kiple 2007). In 1539, several hundred came ashore with de Soto and his men near Tampa, Florida. For four years this group of adventurers marched ruthlessly through a vast territory that is now Florida, Georgia, Mississippi, Tennessee, North and South Carolina, Arkansas, and Texas. While searching for gold, they nonchalantly brutalized and killed countless natives.

But according to some historians, the worst thing that de Soto did was to bring the pigs. By all accounts he was a brutal and rapacious man, but in this case he meant no harm. De Soto simply brought the pigs along for food; he had no idea that his walking meat lockers were carrying death in their lungs and in their intestines. Pigs can incubate and spread an array of horrific diseases, including influenza, anthrax, brucellosis, and tuberculosis. Some of the animals escaped during the journey or were released into the surrounding woods. There they reproduced rapidly and began spreading disease organisms to other animals, such as deer and wild turkeys.

Eventually these diseases made their way into the Native American populations, whose immune systems were totally unprepared for such an assault. More than 90 percent of infected populations died, resulting in the destruction of entire cultures. When de Soto's army invaded what is now southern Arkansas, they noted that the area was "very well peopled with large towns... two or three of which were to be seen from one town." But the pigs and their diseases changed that. When the French explorer La Salle visited the same area more than a century later, he found only a few people living there among the ruins of empty towns and villages (Mann 2002).

The once-thriving Caddoan civilization had been destroyed, brought to near-extinction by invisible organisms carried in the bodies of Spanish pigs. Pigs are clearly the villains of this piece, but they were unknowing villains; mass murderers, yes, but innocent mass murderers. In this case, the harm they did was horrific. But elsewhere in America and at a later time, they would play a different role, feeding a growing American nation for nearly three centuries.

The Pilgrims

On a cold November day in 1620, a small battered ship, the Mayflower, anchored off the coast of what is now Massachusetts with 102 forlorn Pilgrims aboard. Bad luck and a series of betrayals had caused them to arrive on an unknown coast near the beginning of a brutal New England winter with their food running out and no shelter. They had much in common with their biblical namesakes (Hebrews 11:13-16).

> [T]hey were strangers and pilgrims on the earth...And truly, if they had been mindful of that country from whence they came out, they might have had the opportunity to have returned. But now they desire a better country...

It is disturbing how easily historical associations can be lost. In my copy of the *New American Standard Bible*, the phrase "strangers and pilgrims" has been changed to "strangers and exiles." But back to the story. At first it was not certain that the Pilgrims had come to a better country. By spring, half of them lay dead beneath the cold New England soil and most of those who survived were sick and dispirited. Historians have marveled not that so many of the Pilgrims died that first winter, but that so many of them somehow managed to survive. The "somehow" consisted of invading deserted Indian villages and stealing the caches of corn left behind by the dead inhabitants.

Governor William Bradford later wrote, "And sure it was God's providence that we found this corn, for else we know not how we should have done" (Mann 2005). The fact that stolen corn helped save half the Pilgrims from starvation and death during their first winter in New England was significant, foreshadowing events to come. Corn would never again be quite as important to our country's survival, but centuries after this first contact it would become the keystone species in America's industrial food system.

A fateful and tragic event had prepared the New England coast for the arrival of the Pilgrims, enabling them to survive by pilfering corn from silent, eerie Indian villages inhabited only by the dead. Several years before our hapless founders reached the coast of America, a French ship had wrecked near Cape Cod and local Indians captured some of the survivors. Tragically, one of the shipwrecked sailors suffered from a deadly disease now believed to have been viral hepatitis. Some of his captors became infected and the disease raged like wildfire among the coastal Indians, ultimately killing as many as 90 percent of them. So instead of casting ashore on a densely-populated and well-defended coast, the Pilgrims found themselves in a region that had been depopulated only a few years before. Even more favorable for them, the former inhabitants had left food behind.

Governor Bradford was again suitably grateful, declaring "The good hand of God favored our beginning [by] ...sweeping away great multitudes of the natives...that he might make room for us" (Mann 2005). New arrivals to the New England coast continued to take advantage of the room that the "good hand of God" had prepared for them. More than 50 of the first colonial settlements in the region were located on the sites of former Indian villages that had been emptied by disease.

In late March, a delegation of Indians led by the local chief Massasoit visited the surviving Pilgrims. An interpreter named Tisquantum accompanied them. He had been captured by a passing ship some years before and carried away to England, where he learned to speak the language fluently. Chief Massasoit was, in his own way, as desperate as the Pilgrims and anxious to make a deal. The epidemic that made room for the Pilgrims had done so by killing off most of his people. Massasoit and the remnants of his tribe were vulnerable, threatened by the Narragansett, their longtime enemies to the west, who had escaped the pestilence. Massasoit neither liked nor trusted these short, dirty, unpleasant white men, but they had guns, and thus an alliance with them just might save his people. Massasoit offered to leave the Pilgrims in peace if, in return, they would help him hold off the Narragansett. They came to an understanding and as part of the arrangement, Tisquantum remained with the Pilgrims.

When I attended elementary school, every child was told the story of how a friendly Indian came to live with the Pilgrims and taught them to grow corn. We were led to believe that he acted out of benevolence, but that is not true; it was all part of a calculated political deal. In addition, neither we nor our teachers understood just how important this arrangement would turn out to be, both for the European invaders and for the Native Americans. Incidentally, Tisquantum (often shortened to Squanto) probably was not his real name; roughly translated, Tisquantum means something like "the wrath of God or wrath of the Great Spirit." If the

Indians could have known what his actions ultimately would lead to, they might have thought the name appropriate.

Corn has a number of advantages as a food crop. Along with a few other plants, corn has evolved a way to limit water loss during photosynthesis. It has other properties that made it invaluable to American pioneers. Corn could grow wherever wheat could grow and even on land that was too wet for wheat. On land of the same quality, corn required less labor to grow and yielded about twice as many calories per acre. A corn farmer could expect to get back about 150 kernels of corn for every one planted; a wheat farmer on the same land could expect a return of only five or so seeds for every one planted. Corn also had superior disease resistance.

Fortunately for the Pilgrims, Tisquantum taught them how to grow, harvest and store this wondrous crop. Some historians believe that, by giving the Europeans this technology, Tisquantum provided them with just what they needed to eventually destroy Indian civilization in North America. According to one author, "Corn was the means that permitted successive waves of pioneers to settle new territories. Once the settlers had fully grasped the secrets and potential of corn, they no longer needed the Native Americans" (Warman 2003). It is ironic that this sacred native crop would feed and enrich the European invaders as they swarmed across the continent to displace the Indians from their ancestral lands.

The Great Eastern Forest

Nathaniel Hawthorne's iconic short story *Young Goodman Brown* provides chilling insight into the kind of fear and desolation the first settlers in America must have felt. They found themselves in a strange land, fearfully distant from all they had known, on the edge of a vast, hostile wilderness. Hawthorne tells of a young Puritan man in the 1600s who one night spies his minister, his fellow townspeople, and even his young wife communing with Satan's disciples in a clearing deep within the forest. Hawthorne describes the young man's passage through the dark woods.

> He had taken a dreary road, darkened by all gloomiest trees... It was all as lonely as could be; there is this peculiarity in such solitude, that the traveler knows not who may be concealed...so that with lonely footsteps he may yet be passing through an unseen multitude...he glanced fearfully behind him... "What if the devil himself should be at my very elbow!"

The young man keeps going and soon comes upon a disturbing sight. He sees neighbors and community leaders he has revered since childhood in the midst of sinful revelry.

> [C]onsorting with these grave, reputable, and pious people, these elders of the church, these chaste dames and virgins, ...were men of dissolute lives and women of spotted fame, wretches given over to all mean and filthy vice, and suspected even of horrid crimes...the good shrank not from the wicked, nor were the sinners abashed by the saints. Scattered also among their pale-faced enemies were the Indian priests...who had often scared their native forest with more hideous incantations than any known to English witchcraft.

One can imagine many young men like Goodman Brown during the early days of our country, standing at the edge of the forest near their tiny villages and peering fearfully into the gloom of a dark winter evening. Cooking fires and candles flickering dimly in cabin windows would provide little comfort — puny defenses against the endless darkness. Beyond was wilderness, going on forever, inhabited by bloodthirsty natives, fierce beasts, and worst of all, Satan and his minions. At times fear and superstition would drive whole towns mad, and in their madness, they would hang and burn innocent women as witches. But gradually they became less fearful of their new world. Tentatively at first and then more boldly, succeeding generations moved deeper into the vast forest, began to learn its secrets, and prospered.

We should never forget the importance of geography, the significance of place, in human affairs. What if the Pilgrims, instead of landing in a forested, well-watered land, had come ashore at the edge of vast, rolling grasslands or a dry, sandy desert? Many things would have been different. For one, literary passages such as the ones above might never have been written. It is important to remember that even intellectual pursuits such as literature, religion, and philosophy, which at first glance seem wholly of the mind, are affected greatly by where people live and the nature of their environment. The more practical arts, such as economics and politics, are influenced even more strongly. And as to the topics with which we are mainly concerned, such as the kinds of agriculture practiced and the types of food eaten, words such as influence or affect hardly suffice; control might be more accurate and descriptive.

As they gradually overcame their fears and learned its ways, colonists found a home in the vast American forest. It provided them with an abundance of shelter, warmth, and food. Unlike their European cousins, early colonists did not have to shiver through the winter in cold, poorly-heated rooms. There were plenty of tall, straight trees for building snug cabins and no scarcity of fuel for heating them. Frontier families burned an extraordinary amount of wood, often 20 to 40 cords in a single winter. This is amazing when one remembers that the standard cord is a stack measuring four feet wide, four feet high, and eight feet long. Most people burned more wood to heat their house each year than they had needed to build it initially. An English visitor wrote of America in 1650 that a "poor servant here...may afford to give more wood for timber and fire...than many Noble men in England can afford to do" (MacCleery 1992).

Corn and Pigs

In the 1700s, it was common for settlers moving into wilderness areas such as western Virginia, Pennsylvania, Kentucky, or Ohio to establish a *tomahawk claim*, also called *cabin rights*. The settler would claim ownership by blazing or girdling several trees near a spring or other prominent landmark. Usually, he would inscribe his initials or name on the back of one of the blazed trees. A tomahawk claim gave the settler no legal rights unless he then occupied the land, grew a crop, and secured a patent or warrant from the land office. Tomahawk rights generally were recognized as valid, and many of these settlers eventually gained legal ownership of the land. For example, according to one historian (Wiley 1883),

> Virginia gave to every bona fide settler who built a log cabin and raised a crop of corn before 1778, a title to 400 acres of land and a pre-emption to 1,000 acres more adjoining...[C]ommissioners were appointed to give certificates of these "settlement rights." The certificate with the surveyor's plat was sent to the land office at Richmond, and in six months if no *caveat* was offered, the patent was issued, and the title was complete.

This historian went on to write that, in order to get a certificate of ownership, the applicant had to pay "ten shillings per one hundred acres." The cost of the certificate itself was "two shillings and sixpence."

Note that in order to qualify for land ownership under the "cabin rights" system, a settler had to have raised a corn crop on the land. It is fortuitous that the law specified corn. When Europeans first arrived in America, most of the land between the Atlantic Ocean and the Mississippi River was heavily forested. The heavy tree cover was an impediment to farming. Cutting the trees, uprooting the stumps, and preparing the land to plant a European crop such as wheat would have taken years, and the settlers were working against time. Most of them came into the wilderness with a limited stock of food, so they had to produce a crop quickly or they and their families would starve.

They were able to solve this problem by growing corn and by using the same planting and tilling methods as the natives. Cutting trees to build a cabin created openings in the canopy, allowing light to come through. If necessary, they then used fire to clear out any undergrowth in the cut area, making the site suitable for planting. To establish a crop the next spring, the settler would dig holes spaced about a yard apart and drop four corn seeds into each hole, following the practice of the Native Americans. After the corn had sprouted and begun to grow, a hoe was used to form a small mound of earth around each stalk. After a few years of cultivation, the mounds became well established and were reused each year.

It was common practice to plant the seeds of other native crops, such as beans, squash, and sunflowers, along with the corn. Beans were planted very close to the corn, allowing the vines to grow up the stalks and use them for support, while squash and sunflowers were planted at a greater distance. The practice of mixed farming increased overall yield and reduced competition from weeds.

Mixed farming or intercropping also provided a more varied and healthful diet. For example, corn and beans together contain all of the essential amino acids the human body needs to make protein. Some of the corn and beans were eaten very early in the season while both the bean pods and ears of corn were young and tender. Later in the season, fully mature corn could be cooked and eaten or dried and ground into corn flour. Corn flour was used to make bread or cooked in water to make mush or grits. In most pioneer cabins, a pot of mush cooked constantly over an open hearth fire. Both corn and beans were dried and kept over the winter.

Cultivated foods such as corn, beans, and squash were supplemented by nuts, game, and berries from the surrounding forest and by fish and shellfish from nearby streams. This made for a reliable food base. But to be fully food secure, a frontier family needed something else — a domestic animal to serve as a source of meat and fat. So soon after building a cabin and getting the first crop planted, the next order of business for nearly every settler, unless he had brought some with him, was to buy or trade for some pigs.

Pigs are not native to North America, but nearly every group of Europeans brought a herd with them when they sailed to the new world. In 1623, only three years after its founding, Plymouth colony was home to 50 hogs. A decade later Massachusetts Colony was described as having "swine innumerable" (Harris 1998). The hogs, for their part, could not have fared better; by accompanying Europeans to North America, they had found a place in which they could thrive — a hog's Garden of Eden.

It is fortunate that hogs were so well suited to the humid, eastern half of what would become the United States, for they do not flourish in some parts of the world. They are so ill suited to some regions that several major religions have forbidden their followers to eat the animals — or even to touch them. Both the Jewish and Muslim holy books describe hogs as filthy and abominable. The Old Testament gives specific instructions, "Of their flesh you shall not eat, and their carcasses you shall not touch; they are unclean to you...everyone who touches them shall be unclean." Muslims are similarly warned in the Koran, "These things only has He forbidden you: carrion, blood and the flesh of swine."

To non-believers these pronouncements might seem arbitrary and unfair, but they reflect certain realities; if forced to live in the wrong climate, notably one that is hot and dry, hogs can have serious personal hygiene problems. The Middle East, ancestral home to both Islam and Judaism, is an example. Since hogs lack functioning sweat glands, they have difficulty getting rid of body heat. This problem is compounded by the fact that they lack fur; only sparse bristles cover their skin, providing little protection from the sun. To avoid overheating, hogs need both shade and mud to wallow in. If it gets much hotter than 85 degrees Fahrenheit, hogs deprived of moist earth will take desperate measures; they will create mud using their own urine and feces. It is easy to see why people living in hot, dry climates would prefer not having them around.

But pig hygiene would not be a problem in the New World. From the Atlantic Coast to the Mississippi River and a little beyond was prime pig country. This thickly-forested, well-watered area provided abundant shade and innumerable mud-wallowing sites for the occasional cool down. It also provided an almost unbelievable abundance of food. Take even a short walk through any forested area in the eastern United States and you undoubtedly will see several varieties of acorn-bearing oaks (*Quercus*) and nut-bearing hickories (*Carya*). There are a few areas where such species are absent, but much of the Midwest, Atlantic coast, and South were dominated by oak–hickory forests when Europeans first arrived here. These primeval forests were home to more than 20 species of oaks and to more than ten species of hickory. Other nut-bearing trees included the now-extinct chestnut, the butternut, the pecan, and the hazelnut (Walker 1998).

Biologists classify hogs as "opportunistic omnivores" and that says it well. In addition to acorns, which make up a large part of their diet in the wild, hogs eat a great variety of other foods, both plant and animal. In the spring they might browse on the new leaves and stems of succulent forbs or, if rains have done a good job of moistening the forest floor, feast on newly-emerging mushrooms. Much of their food is taken out of the ground. Hogs are well-known for rooting, or tearing up the earth with their snouts.

Small soil-dwelling creatures such as insects, grubs, and worms make up a significant part of their menu throughout the year. In addition to eating nearly every snake they come across, hogs are known to kill and eat fawns, lambs, and in some cases full-grown deer, sheep, and goats, especially if the animal is injured or weak. Hogs are clever, aggressive foragers and will eat practically anything, dead or alive. As a result, they can thrive in almost any forested environment where the ground is not frozen or covered with snow for long periods (Gade 2000).

Settlers had only to clear the forest of wolves and hardy "wood pigs" could go forth and multiply. It was fairly easy to lure most of them back to the homestead at butchering time, usually in the late fall. In colder areas such as New England, hogs foraged in the woods only during warmer parts of the year, and were penned and fed during the winter. Farther south they rooted in the forest floor for their living year-around. Farmers soon found that it paid to feed pigs on corn for a month or so before slaughter in order to fatten them up and make their flesh firmer. This procedure of "finishing" hogs on corn before butchering became a widespread practice as early as 1700.

Hogs are very efficient at turning whatever they eat into tasty meat. During its lifetime a modern hog can gain up to a pound of weight for every three pounds of food it eats, a conversion rate much greater than that of cattle. Part of the hog's advantage comes from an unusually low rate of basal metabolism, about 40 percent lower than would be predicted by its body weight. This allows pigs to channel a much higher percentage of consumed food into weight gain.

They have reproductive advantages as well. Within "three months, three weeks and three days" as one grizzled old hog farmer put it, a sow can give birth to eight or more piglets, each of which can gain as much as 400 pounds within the next six months. The enterprising sow can perform the same feat the following year, then again the year after that and so on — as can her many female offspring (Harris 1998).

Pork soon became the meat of choice in colonial America. In 1726, a party of men was dispatched to survey the boundary line between Virginia and North Carolina. William Byrd later published an account of this journey, a *History of the Dividing Line Betwixt Virginia and North Carolina*. In it, he described a section of North Carolina, which would have equally applied to many other farming communities of the time:

> The only business here is the raising of hogs, which is managed with the least trouble, and affords the diet they are most fond of. The truth of it is, the inhabitants of North Carolina eat so much of swine's flesh, that it fills them full of gross humours [they smell bad]...Thus, considering the foul and pernicious effects of eating swine's flesh in a hot country, it was wisely forbidden and made an abomination to the Jews, who lived much in the same latitude as Carolina.

Despite the disgust of gentlemen such as William Byrd, colonial farmers embraced the raising and eating of hogs with enthusiasm. A foreign visitor noted that when one used the word food in Europe, he usually meant bread, whereas in the American colonies the word food usually meant pork. This was reflected in the literature of the time, with one character in a James Fenimore Cooper novel declaring, "Give me the children that's raised on good sound pork afore all the game in the country...pork is the staff of life." Most early Americans agreed; by 1860, there were two pigs for every man, woman and child living in the South and Midwest. A Georgia doctor, appalled that his patients and fellow-citizens ate nothing but bacon or ham at each and every meal, proposed that America be renamed "The Great Hog-Eating Confederacy" (Root and de Rochemont 1976).

Historians have made much of the first meetings between English settlers and local Indian tribes, both at Plymouth, Massachusetts and at Jamestown, Virginia, hundreds of miles to the south. But the meeting between hogs and corn, although largely uncelebrated, might have been equally significant. Together, they became the keystone food species in early America, and their influence was not short-lived. They would dominate American agriculture and together provide most of our calories and protein for the next 300 years. Some historians estimate that, in the absence of corn and hogs, it might have taken a century longer to settle the land between the Atlantic Coast and the Mississippi River.

As settlers moved across the eastern mountains and into the Midwest, hogs and corn moved with them. It was there that pig raising became truly dependent on corn. Farmers were able to grow large surpluses on the fertile Midwestern soils. They soon quit raising "wood pigs" and switched over to breeds which were larger and produced more fat. These "lard pigs" did not forage in the woods as their bumptious cousins had done; instead they stayed in sturdy pens and were fed year around, mostly on corn. At first farmers gathered up their pigs and herded them cross-country to be sold. But this mode of transport soon changed. Instead of having to walk, hogs were butchered at central locations and shipped to market nestled in barrels of salty brine. Bacteria or other microbes could not grow in the bitter brine, so the barreled pork stayed edible for a long time and could be shipped great distances.

When considering the dietary history of America, the importance of pork cannot be overstated. Until the end of the Civil War, Americans ate more of it than any other food except bread, and more pork was eaten than any other meat well into the 20th century (Bondi 1982). In the 1800s, a family in dire financial straits was often described as "looking at the bottom of the pork barrel." As the nation grew, pork became increasingly important to commerce and to economic well-being. Pork and prosperity became so intertwined in people's minds that the phrase "pork barrel" came into usage in reference to wasteful government spending. Even today, legislators are either praised or castigated for their skill at "bringing home the pork." Our country owes a debt of gratitude to hogs and to the rich forest and abundant corn on which they fed. For nearly 300 years after the Pilgrims landed, these resources nourished a rapidly growing population, enabling each successive generation to become more numerous, taller and more robust.

The Bluegill Effect

Many years ago I rented a garage apartment from a widow who lived in the country. She had a small pond out back that had been stocked with fish — bluegills. Bluegills are native throughout much of North America east of the Rocky Mountains and are commonly released into rivers, lakes, and ponds. They provide excellent fishing, especially for children, who like their pecking or nibbling style of feeding and the way they often steal bait from their hooks. Almost any kind of bait can be used to catch them, including live insects or worms, chunks of hot dog, raw chicken, pieces of bread, and corn kernels.

My landlady would not allow anyone to fish in her small pond. As a result, the bluegill population had exploded. I often would walk up to the pond in the evening and throw bread crumbs on the surface. Since bluegills are unusually sensitive to disturbances of the water, they immediately would rush to the surface. The pond seemed to boil with their frenzied activity, at times appearing to be half water and half fish. Another thing struck me; the fish were diminutive, unusually small for bluegills. It was clear what had happened; with no natural predators and no anglers to thin their ranks, the fish population was booming. In order to accommodate such large numbers in a small pond, individuals could not reach their optimal growth and became stunted.

For several centuries before and after the settling of America, Europeans were subject to a similar process, what could be called the "bluegill effect." Like the fish in my former landlady's pond, their numbers increased so rapidly that the land and food crops needed to support them could not keep up. The population of every European country at least doubled from 1600 to 1900, with some growing even more rapidly. During those three centuries, the population of France increased from 20 to more than 40 million, while that of Italy nearly tripled, going from 12 to 32 million. During the same 300 years, the population of Great Britain and Ireland grew from a little under five million to more 40 million.

But these growing populations were not well-fed, happy people. The mass of European and British humanity was malnourished, frequently hungry, and bowed down by hard work. In a few generations these stressed populations, especially those living in crowded, squalid cities, began to exhibit the "bluegill effect" — to become short and stunted. Except for a fortunate few descended from the well-fed nobility, nearly everyone was short. But it seemed the natural order of things. A British general of the time reportedly remarked that if one had to go to war, the ideal infantryman would be a stalwart Welshman five feet, two inches in height.

The first Europeans to arrive in North America were surprised at the height and overall healthy appearance of the Indians they met. In nearly every case, Native Americans towered above the newcomers. A Spanish explorer in 1528 described the Indians in Florida whom he met in battle as follows: "All the Indians we saw...are archers, and because they are naked and so tall, from a distance they seem gigantic. They are a people admirably well-formed, very slender, and with great strength and quickness" (Howard 1997). William Wood expressed similar admiration for the Indians of New England, writing that they were "more amiable to behold... though only in Adam's finery...than many a compounded fantastic [English dandy] in the newest fashion." Italian explorer Giovanni da Verrazzano observed that a Narragansett chief who came aboard his ship off the coast of New England in 1523 was as "beautiful of stature and build as I can possibly describe" (Mann 2005).

It is little wonder that the Indians of eastern North America appeared so tall and physically imposing to the Europeans. They had only recently adopted agriculture and had not yet given up hunting and gathering. Much of their diet still came from the rich natural bounty of the complex forest, ocean shore, marsh, and riverine environments that they were fortunate to inhabit. An abundance of fish, shellfish, birds, upland game, nuts, and fruit were free for the taking. This healthy and varied diet was made even richer by the annual harvest from productive fields of corn, squash, and beans grown in fertile river valleys. Their lifestyle was an active one, but they were not bowed down by hard labor.

Captain John Smith, who famously recounted his last-minute rescue by the young Indian princess Pocahontas, later sailed along the coast of Massachusetts. His journey was in 1614, shortly before the area was devastated by disease. Smith was very taken with the country, noting that the land was "so planted with Gardens and Corne fields, and so well inhabited with a goodly, strong and well-proportioned people [that] I would rather live here than anywhere" (Mann 2002).

It is doubtful that the European immigrants realized how fortunate they were in coming to such a place. To help us understand how lucky they were, let us return to the story of my former landlady's fish. Assume that I had scooped up ten or 15 of them from their small pond and later released them into a vast, pristine freshwater lake. If their fishy little brains could understand their good fortune, they would have been overjoyed. In any event, they would have begun to reproduce happily, filling up the new territory open to them. In addition to becoming more numerous, future generations would grow larger, soon reaching normal size for the bluegill species.

European emigrants to America found themselves in just this situation. Most of them were short when they arrived here, not more than an inch or two over five feet in height. Centuries of calorie and protein deprivation had bred a stunted, but tough and resilient race. But like our hypothetical bluegills, they (or rather, their descendants) began to change, spreading out across a new continent and multiplying rapidly. At the same time, with an abundance of nutritious food available to them, successive generations began to grow taller.

In 1620, not more than a few hundred Europeans lived in the vast expanse of land that was to become the United States. This included the inhabitants of New England and Virginia as well as Spanish priests and soldiers of fortune in Florida and the southwest. But, through a combination of reproduction and immigration, their numbers soon began to increase rapidly. The United States government conducted a national census shortly after it came into existence. This census, taken in 1790, showed that the population of the new republic was a little less than four million. The 1850 census recorded more than 23 million Americans. By the time of the 1900 census

50 years later, the population had more than tripled, reaching nearly 75 million (U. S. Bureau of the Census 1975).

As Americans grew more numerous, they also became taller. During the Revolutionary War, the average American soldier was a little more than five feet, five inches tall. He was short by modern standards, but still several inches taller than the English conscripts he was to face in battle. Less than a century later, military officials measured the heights of more than 500,000 Civil War soldiers, mostly young men from old families. The phrase "old families" refers to those that had been in this country for two generations or more. These men, at an average height of nearly five feet, eight inches, were more than two inches taller than their Revolutionary War counterparts had been less than ninety years before.

In 1866, for reasons that are not clear, government officials measured the heights of all members of the United States Senate. The results showed a pronounced "old family" effect. These men, almost without exception, were descended from successful families that had been in this country for many generations. Measured without their shoes, they averaged nearly five feet, ten inches in height, exceeding, according to the report, "the average of mankind in all parts of the world as well as the average of our own country." New immigrants to America continued to follow the pattern illustrated by the "old family" studies. Many would arrive short in stature but live to see their children and grandchildren become much taller, a process that continues even today (Hathaway 1959).

Although each generation of Europeans grew taller on the high protein diet available to them in their new country, they did not grow fat from this abundance. Instead, generation after generation remained lean, gaining weight only in relation to the extra muscle and bone needed to accommodate their taller bodies. The population was kept lean by a combination of moderate to strenuous work and a diet consisting mostly of natural foods. By 1880, two out of three Americans still lived on farms, and one in four still tilled the soil as late as 1930. Most town and city dwellers were but one generation removed from the land. Before the 20th century, people living on farms and in small towns baked their own bread, gathered vegetables and fruits from their own gardens and orchards, and ate the meat of lean, active, often free-range animals. They spent part of each day outside doing chores in the sunshine and fresh air. Nearly everyone had to work hard, but few were broken down by toil. The truly obese person was a rare oddity before 1900.

For most of American history, the dietary news was mostly good, but in the 20th century, especially in the second half of it, things began to go bad. American families continued to get taller with each generation, edging closer to their biological potential. But something unpleasant and unexpected also began to happen; each generation started to become fatter, slowly at first and then at an accelerated rate. Now, in the early years of the 21st century, we are confronting an epidemic of obesity and diabetes that threatens to rage out of control. Increasingly, physicians are saying that we must take drugs or undergo radical surgery to keep our food from killing us.

The Corn Children

On the day I first wrote the paragraph above, I was sitting in a fast food restaurant with one of my granddaughters. She was enjoying what to her was a gourmet meal — chicken nuggets, fries, and a coke. I remember glancing around the dining room full of young suburban mothers and their charges. Nearly all of the children were eagerly consuming the same culinary delights as my granddaughter. Shredded and reconstituted chicken was definitely the food of choice among the younger set.

Some form of pig meat was the preferred food in America for most of our history. But sadly, pork was dethroned as the king of meat shortly after World War II, when beef became the favorite. But beef was to have only a brief reign; Americans recently embraced chicken and now eat more of it than any other meat, much of it in the form of "nuggets."

The glory days of the hog are over, but what about corn, his tall, elegant, golden-tasseled partner? How did she fare in the 20th century? The simple answer is that corn did very well indeed. To confirm this, you need only consider the meals my granddaughter and the other children were eating. The chicken nuggets they were devouring so avidly were corn to the core. A batter made from fine corn flour coated each delectable tidbit, giving it an acceptable non-animal, non-chicken look. Even the meat itself came mostly from corn; an anonymous chicken had to eat about half a pound of corn grain to produce the six nuggets each child got as part of his or her "chicken nugget value meal." The children were even drinking corn in the form of high fructose corn syrup that sweetens their Cokes.

It is no exaggeration, though perhaps a pun, to say that the kids around me were definitely corny. If you took cuttings from their fingernails, from their hair, or for that matter, any of their tissues and ran them through the appropriate laboratory instrument, you would find that most of the carbon atoms in their bodies got there by way of a corn plant (Pollan 2006). You might think that this "cornification" of America is a very recent thing, but it is not. Although it might have accelerated in recent years, it actually started almost a century ago. Consider the following passage (Dies 1949).

> Today [1949] no other single agricultural crop makes up so large a part of the American diet, directly or indirectly, in the form of meat, milk and eggs...[Y]ou are getting products of corn for breakfast, lunch, and dinner...But corn's versatility extends far beyond its primary use as food and feed. It passes into an amazingly long list of industrial products...

The passage above, written in 1949, is even truer today. Walking down the aisles of a modern supermarket is, in a way, like walking through a corn field. More than one in four of the thousands of items on display will contain corn. It has found its way into such things as toothpaste, cosmetics, baby powder, trash bags, disposable diapers, and charcoal briquettes, even the wax on cucumbers. As for food, it is almost impossible to find any packaged or canned item that does not contain high fructose corn syrup. Check the contents label the next time you buy bread, cookies, ketchup, sweet pickles, spaghetti sauce, or canned soup. Chances are you will find it.

There is something else interesting about my granddaughter and the other children described above. Depending on their weight, each of them has about a pound of nitrogen in his or her body. The fact that the nitrogen is there is not the interesting part; what is fascinating is where it came from. Up to half of the nitrogen in each of these little bodies is not "natural," but is industrial in origin. It came from one of relatively few highly-efficient ammonia factories located around the United States, often in places such as Texas, Oklahoma, or Louisiana to be near natural gas sources.

These factories are designed to do just one simple thing. They pass a heated, pressurized mixture of nitrogen (N_2) from the air and hydrogen (H_2), usually from natural gas, over a catalyst. These extreme conditions cause the nitrogen and hydrogen to combine, forming ammonia, or NH_3. This valuable end product is made into fertilizer and applied to the soil, where the added nitrogen greatly increases plant growth and crop yield. More nitrogen fertilizer is added to corn than any other crop. The plants absorb the factory-made nitrogen from the soil and use it to make corn protein.

Eventually, little girls and boys eat the corn, usually after it has been eaten by a chicken or a steer and then processed into a chicken nugget or hamburger. In this industrial system, an atom of nitrogen might make its way from a factory in east Texas into a corn plant in Iowa, and then into the body of a chicken in Arkansas, then perhaps take up residence in the arm of my young granddaughter here in North Carolina, forming the muscles that enable her to reach out and pick up yet another chicken nugget.

Sitting there in that restaurant, we were at one of the final delivery points of a vast industrial food system, a 20th century phenomenon. The basic elements that make such a system possible are fossil fuels, hybrid corn, and industrial nitrogen. This is the first time I have used the word "hybrid" in reference to corn. All commercially grown corn in the United States is the product of a rather bizarre hybridization process discovered around 1900.

You might think that such an obscure, dull-sounding topic could be of interest only to farmers, agricultural agents, or people in agribusiness. But the reality is this: in the second half of the 20th century, hybrid corn partnered with industrial nitrogen to transform our agriculture, our food, and our very bodies. They made the age of industrial food possible, changing how we eat, how we live, and ultimately, how most of us will die. Although hybrid corn and industrial nitrogen have been a boon to humankind in many ways, their dark side is proving to be very dark indeed. They are major players in an ongoing dietary and public health drama that opened in the early part of the 20th century and is now working its way toward a seemingly inevitable and perhaps tragic end.

Chapter 7. Industrial Food

In 1920, America was home to 100 million people. At present, as of 2021, our population has grown to nearly 330 million — having more than tripled in less than a century. Despite an astounding increase in the number of people to be fed, America's farmland acreage has not increased at all; in fact, it has declined slightly. America now feeds a population three times larger than in 1920, and each of us eats more meat, dairy products, fat, calories, and sugar than our grandparents did, while using slightly less farm land. At the same time, America has become a major exporter of many farm products. This chapter describes two scientific breakthroughs most responsible for making that possible — hybrid corn and industrial nitrogen. Together, these two 20th century innovations changed how we live and what we eat. They did much to make us into a more urban society, they helped us to become richer and more numerous, and now they are making us fatter and sicker.

Nitrogen

Near the end of Chapter 5, we described Sir William Crookes' 1898 presidential address to the British Association for the Advancement of Science, in which he warned that England and other "civilized" countries faced starvation and civil unrest as early as 1930 unless something were done soon. Crookes went on to specify who should take action. "It is the chemist who must come to the rescue... It is through the laboratory that starvation may ultimately be turned into plenty."

Although nitrogen makes up most of the atmosphere, getting enough of it into our bodies has been a problem throughout human history. The potential supply is virtually unlimited, but in order to be biologically useful (available to build protein), nitrogen must be in the right form. Because of its perverse chemistry, most of the human race has been nitrogen- and, as a result, protein-deficient for many centuries. Short, stunted physiques have been the defining characteristic of agricultural populations worldwide.

Nitrogen in the air does not readily combine with other elements. Instead, it much "prefers" to exist as highly inert nitrogen (N_2) molecules held together by very strong triple bonds ($N{\equiv}N$). Nitrogen cannot be used to build plant and animal protein until those bonds are broken. Two natural entities on Earth are capable of doing this. One is lightning and the other is an enzyme called *nitrogenase*, found in the bodies of a very small number of microbes.

As many as eight million bolts of lightning strike the earth each day and millions more zigzag across the sky without ever reaching the ground. Each searing flash produces unimaginable heat in only a few millionths of a second, creating plasma that is hotter than the surface of the sun. This immense surge of heat and energy breaks apart many of the $N{\equiv}N$ bonds. Some of the freed nitrogen quickly combines with oxygen to form NO and then NO_2. The NO_2 readily dissolves in rainwater and is washed out of the air and deposited on the earth's surface. Additional rain washes it into the soil, where it can be taken up by plant roots. Removing nitrogen from the air and converting it into a form that plants can use is called "nitrogen fixation." Each year lightning enriches each acre of land with a small amount of nitrogen, helping make it possible for the land to clothe itself in green vegetation.

Lightning is not alone in being able to take nitrogen out of the air and convert it into a form plants can use. Early farmers observed that certain kinds of plants, such as beans, clover, and alfalfa, seemed to enrich the soils in which they grew. Although they had no way of knowing it, these early farmers were witnessing the results of a complex biological process that was taking place underground. A small number of plants have the ability to take nitrogen out of the air by forming an intimate partnership with soil microbes which take up residence in their roots.

The microbes start out as tiny spheres, among the smallest organisms on earth. After changing their form and growing a "tail," they move through the soil until they approach the root of a receptive plant. The root, upon "sensing" their presence, reacts by secreting a substance that causes the strange little creature to swim toward it. The microbe then releases a substance into the soil that causes the plant root to bend. It enters the root at that spot, settles in and begins to multiply, soon forming a visible nodule. A partnership has been formed. The plant provides the microbes with shelter, water and food and in return, the little visitors take nitrogen out of the air in a form the plant can use to grow (Russell 1971). In the aggregate, microbes fix much more nitrogen than lightning.

When legumes such as soybeans and clover die and decompose, the nitrogen stored in their tissues is released into the soil, acting as a natural fertilizer. The thrifty earth hoards this valuable nutrient by stashing it away in the upper layers of the soil. In natural systems, very little nitrogen is lost to surrounding streams, rivers, or lakes. Over time, large quantities can build up in soil organic matter, and each year a small part of it is made available for plant growth. This usually happens in the spring when soil fungi and bacteria "wake up" from their winter sleep and start breaking down organic matter for food — releasing nitrogen in the process.

Although they had no idea of what was going on, both Chinese and Roman writers urged farmers to plant nitrogen-fixing crops as early as 500 BC. As time went on, more and more people learned to take advantage of this gift from nature. Early farmers soon learned another trick — that they could replenish the soil's nitrogen supply by carrying in nitrogen-rich organic material from other areas and putting it on or in the soil. Thus began the practice, still used today, of enriching the land with animal manure, litter from barns, or decomposed crop residues. This became standard practice in Roman and European agriculture, as well as in China and other parts of Asia. The combination of adding organic matter to the soil and using nitrogen-fixing plants made the land much more productive and allowed populations to increase dramatically.

But despite all of the ingenuity humans could bring to bear, population growth always seemed to stay ahead of crop production. Many generations of gaunt, stunted peasants are mute testimony to this. The situation started to become critical in the mid-1800s, at which time England and Europe had been experiencing a population boom for at least two centuries. The populations of Spain and Italy nearly doubled from 1700 to 1800. Other European countries experienced similar population growth. Although they continued to increase in number, Europeans were not well fed and were growing increasingly desperate. By the mid-1800s real trouble was in sight. Crop yields were declining and little new land was left to bring under the plow. A solution had to be found or the developed world soon would face the specters of famine and civil unrest. Fortunately, a gifted and troubled chemist came to the rescue.

It is surprising that Hollywood, to my knowledge, has never made a movie of Fritz Haber's life. His accomplishments, misdeeds, and personal suffering would make for riveting drama. A German of Jewish descent, he was a brilliant chemist. He also was an ardent patriot willing to do anything to serve his country. During World War I, he led German efforts to utilize poisons such as phosgene and mustard gas as battlefield weapons. It is reported that he personally supervised the use of poison gas at the battle of Ypres, in which more than 10,000 allied soldiers were killed in a matter of minutes. Not surprisingly, many refer to him as the father of chemical warfare.

His wife, also a chemist, was horrified by his activities. Some have speculated that this revulsion led her to commit suicide in the family living room by shooting herself in the heart. According to reports, Haber took little time to mourn. Soon after her death, he left for the eastern front to explore even more ingenious ways to kill the enemies of his beloved Germany with poison gases. Despite committing wartime atrocities, Fritz Haber was awarded the Nobel Prize in 1918. The prize was presented on June 1, 1920, by A.G. Ekstrand, President of the Swedish Academy of Sciences. Following are excerpts from Ekstrand's introductory speech, available on the Nobel Prize website.

> The Royal Swedish Academy of Sciences has decided to confer the Nobel Prize in Chemistry for 1918 upon the Director of the Kaiser Wilhelm Institute, Professor Doctor Fritz Haber, for his method of synthesizing ammonia from its elements, nitrogen and hydrogen... We congratulate you on this triumph in the service of your country and the whole of humanity.

Fritz Haber received this award for a scientific breakthrough he reported in a letter dated July 3, 1909. Below is part of a letter he sent to the directors of his company in which he described the historic experiment he had carried out the day before (Smil 2001a).

> Yesterday we began operating the large ammonia apparatus...and were able to keep its uninterrupted production for about five hours. During this whole time it...functioned correctly and it produced continuously liquid ammonia. Because of the lateness of the hour, and as we all were tired, we...stopped the production because nothing new could be learned from continuing the experiment...

Carl Bosch, a colleague of Haber and a gifted engineer, was given the task of taking what had worked in the laboratory and turning it into an industrial process. Bosch succeeded and soon Germany was producing ammonia in large quantities. Ammonia, or NH_3, is readily converted to ammonium nitrate or other compounds which can be used to fertilize the soil and greatly increase food production.

But nitrogen, especially in the form of ammonia, has other uses. A line from an old nursery rhyme captures the element's essence, "When she was good she was very, very good, but when she was bad she was horrid." In one of its "good" personas, as fertilizer, nitrogen can turn the landscape green and feed the hungry. But at the time, Haber and his employers were not primarily concerned with the good potential of nitrogen. Their main interest was in preserving Germany's capacity to wage war; and nitrogen, in one of its horrid personas, is a strategic war material. Instead of making fertilizer, one can just as readily convert NH_3 to NO_3, which then can be made into a veritable witch's brew of explosives. So, in addition to feeding millions of hungry people in the modern world, Haber's discovery helped make the 20th century the bloodiest in history.

Many did not believe Haber should have been given the Nobel Prize. Instead, they argued that he should have been tried as a war criminal. But Haber was never charged. Although there was much evidence against him, world leaders were not willing to indict a Nobel Prize winner. Nevertheless, difficult times lay ahead for him. The Nazis came to power in the 1930s and the German nation began its steep descent into tyranny. Despite his fame and the sacrifices he had made for their nation, the Nazis viewed Haber simply as another Jew to be despised and distrusted. He became alarmed when officials began arresting and deporting his Jewish colleagues. Soon he fled Germany and sought refuge in Switzerland, where he later died, never returning to the once beloved country that had betrayed him (Smil 2001b).

Fritz Haber remains an enigma — a complex and driven man whose life's work produced so much good while at the same time ushering in a century of death and destruction. But one thing is undeniable; the successful experiment he carried out on July 2, 1909, was a watershed event in human history. Haber's discovery of how to take nitrogen out of the air and turn it into fertilizer was a scientific and historic milestone. This achievement helped make both the huge population growth and the destructive wars of the 20th century possible. Haber also bears some responsibility for the worldwide epidemic of obesity and diabetes that now rages around us. His mixed legacy will continue to haunt us well into the 21st century and beyond.

Hybrid Corn

During the time that Fritz Haber was working on industrial nitrogen in Germany, George Schull, a brilliant and obsessive American scientist, was busy unraveling the mysteries of corn. His goal was to find varieties that would produce more food per acre. In the end, it was not a matter of finding but of creating. A combination of brilliant insight and much trial and error led Schull to the development of hybrid corn. This breakthrough, reported in 1908, would help to revolutionize agriculture and profoundly change the American way of life.

These two discoveries — industrial nitrogen and hybrid corn, have much in common. Both were milestone events in the 20th century, yet few Americans have heard of them or understand their significance. This is not surprising. Such quaint-sounding topics as "hybrid corn" and "industrial nitrogen" have little appeal to a sophisticated, post-industrial society such as ours. But to a considerable degree, the fact that there are so many people on Earth and that so many of them are obese and sick is the direct result of a "marriage" between hybrid corn and industrial nitrogen that took place shortly after World War II.

I remember my first real contact with hybrid corn. Driving down a long, straight rural road in Indiana many years ago, I became intrigued by the veritable walls of corn on either side of me. The plants stood stiffly at attention like well-disciplined troops, uniform in height and so close together that the leaves of adjacent plants rustled against each other in the light wind. I later learned that as many as 30,000 plants grew on each acre. But what really astounded me were the ears. They began growing at almost exactly the same height above the ground and looked amazingly thick and full. I remember thinking, "I could fire a rifle down any row and, as long as the bullet kept going straight, I would be able to pierce an ear of corn on every plant in the row." After arriving home, I went to the library and read up on hybrid corn — this was before the days of Google. One interesting point stood out: Their geometric precision and uniformity came from the fact that all of the corn plants in the field were genetically identical.

The essence of hybrid corn is uniformity, high yield, and adaptation to mechanized farming. A harvester chugging through an Indiana corn field on a brisk fall day is much like a factory machine manipulating highly-engineered parts on an industrial assembly line. Corn has been transformed from a complex biological entity into a simplified unit of industrial production. In the process it has become, in the words of one author, "something never before seen in the plant world: a form of intellectual property...the biological equivalent of a patent (Pollan 2006).

To fully understand such statements, it will help to review the clever, intuitive way in which George Schull created the first hybrid corn. Working at his research station on Long Island, New York, in the early 1900s, Schull began by picking a corn plant and hand-pollinating it with its own pollen. He then planted the seeds from this plant and, when they grew to maturity, went through the same self-pollination process again. The next generation also was self-pollinated, as was the next, for a total of four or more generations. A biographer of Schull described this process as repeatedly "marrying corn to itself" (Dies 1949).

These highly inbred strains became increasingly uniform, but they also lost vigor, becoming spindly, low-yielding, and difficult to propagate. But when Schull crossed two of the sickly, highly inbred lines, the resulting offspring were vigorous and high yielding. This phenomenon, called "hybrid vigor," resulted in initial yield increases of 20 to 40 percent or more, but there was a downside, at least for the farmer. Hybrid vigor did not carry over into the next generation; one could not save seed from this year's crop and plant it the following spring. The farmer's loss was the seed seller's bonanza. In order to keep getting high yields, the farmer had to come back each year and buy more hybrid seeds.

Knowing that they would have the biological equivalent of a patent on anything developed, seed companies were eager to begin producing and marketing new strains of hybrid corn. Farmers quickly became convinced of its value; the huge improvements in yield could not be ignored. By 1940 more than 25 million acres were planted to the new wonder crop, mostly in the Midwest, but the new technology soon spread into the southeast and other corn-producing areas. By the 1950s farmers were harvesting as much as 80 bushels per acre, more than double the 25 bushels per acre or that had been possible during the 1930s. But this was nothing compared to what would happen in the second half of the 20th century.

The Leftovers of World War II

A lot of explosive power was needed to defeat the German and Japanese war machines in World War II, and ammonium nitrate was the main ingredient in those explosives. The United States built its first wartime ammonia plant in 1941. By 1945, we had constructed ten additional plants and were turning out nearly 900,000 tons of ammonium nitrate each year. But then hostilities ended and the nation found itself with a large surplus of this material. What should be done with this stockpiled nitrate and with the infrastructure that had been created to produce it?

Some suggested spraying it on forest land to improve timber growth. But the Department of Agriculture came up with what they thought was a better idea: use the leftover ammonium nitrate and the surplus production capacity to fertilize the nation's food crops. The turning point was 1947. This was when the government started converting wartime munitions plants into fertilizer factories, beginning with a sprawling facility at Muscle Shoals, Alabama. Much that we now take for granted, such as immense, brightly lit grocery stores open 24 hours a day, and fast-food restaurants seemingly on every corner, can be connected directly to that little-noticed event in northern Alabama (Smil 2001b).

From 1950 to 2000, the productivity of America's farms and ranches increased by more than 150 percent. At the same time, the output per person increased so much that only one-fourth as many farmers and farm workers were needed. Furthermore, both the capital investment and the amount of land needed for farming declined during this period. We have several developments to thank for this. First, shortly after the end of World War II, tractors replaced the remaining horses and mules on American farms. This was followed by the use of more fertilizer, especially nitrogen, and the introduction of new technology to apply it more efficiently.

More potent herbicides and pesticides, as well as genetically enhanced plants and animals, soon were added to the mix. Yields increased dramatically for almost all crops. From 1950 to 2000, average wheat yields doubled, going from about 20 bushels per acre to more than 40. The increases in corn yield were even more dramatic, with average yield per acre going from a little over 40 bushels in 1950 to about 130 bushels by 2000. Our best soils now average 150 to 180 bushels of corn per acre, and some farmers even exceed 200 bushels per acre in good years. American farmers now produce about 40 bushels of corn each year for every man, woman, and child in the nation. But the soil needs a lot of help to do this, and most of that help comes in the form of nitrogen fertilizer; more of it is used on corn by far than on any other crop (Doyle 2007).

The Age of Industrial Food

In 1894, Congress appropriated funds to study the nutritive value of the nation's food supply, especially those foods eaten by the working class. The main objective was to develop "wholesome, and edible rations less wasteful and more economical than those in common use" (Todhunter

1959). This legislation came about because of a well-intentioned desire to feed the working poor better in order to make them more productive.

This task was assigned to the U.S. Department of Agriculture and Dr. W.O. Atwater was selected to head the effort. The W.O. stood for Wilbur Owen, but Atwater seemed to prefer the initials. I do not recall seeing his full name in any official correspondence. Atwater was born in Johnsburg, New York, in 1844, the son of a Methodist minister who spent his career moving himself and his family from one small rural community to another. Early photos of Atwater reveal a short, stocky man with a large head and intense dark eyes beneath shaggy brows. Despite being from a poor family, Atwater was able to enter Yale University, where he earned a Ph. D degree in only one year. He then spent two years in Germany studying under some of the world's leading chemists. While there he became fascinated by the emerging field of calorimetry, especially its use in determining the energy content of different foods.

Upon returning to the United States, he spent years conducting laborious, painstaking studies to determine how much energy the human body could capture from the different constituents of food. The figures he arrived at — nine kilocalories from each gram of fat and four kilocalories from each gram of carbohydrate or protein — have not been improved upon in more than a century. A dedicated researcher, Atwater worked furiously until felled by a massive stroke at the age of 60. He survived in a state of near-helplessness for three more years, nursed at home by his wife and daughter until his death in 1907.

Atwater accomplished much during his lifetime and is justly considered the father of modern nutrition research. More than a century after his death, USDA still funds an annual W.O. Atwater Memorial Lecture to recognize scientists who have made outstanding contributions to the field of nutrition. He deserves this honor, but like so many brilliant and obsessive people, Atwater had a dark side. He seems to have inherited some of his minister-father's messianic zeal. While absorbed in his studies of food energy, he became fixated on the dietary follies of the poor. His work had convinced him that food was valuable only as a source of protein and energy, and things such as fruits and vegetables were mere luxuries.

Atwater observed that the poor were paying half or more of their income for food; he felt that they should stop wasting money on such frivolous things as green vegetables and fruits and instead feed their families cheap, energy-rich foods such as flour, corn meal, and rice. He argued that their own dietary indulgences kept them impoverished and ill-housed. According to Atwater, "If the... present waste of food material could be spent for more adequate shelter... the slums would be renovated" (Dies 1949).

Atwater and many of the chemists and nutritionists who followed him valued food only for its chemical components and that influence is felt even today. Few modern nutritionists or diet experts consider food in its totality; instead, they view it only as a useful mixture of fats, protein, and carbohydrates, and in their minds it is these constituents, not the food itself, that really matter. You can see how dominant this reductionist view is by scanning some of the leading nutrition journals. You will find that they hardly mention food at all; instead, they focus almost entirely on the things from which foods are made, such as fats, carbohydrates, proteins, vitamins, minerals, and fiber. The same is true of popular diet books. The next time you are in your local bookstore, pick one up and look through it. Chances are you will see little mention of real, whole foods. Instead, you will be warned against eating too much or too little of some food constituent. Like Humpty-Dumpty, food has been broken into its component pieces and those pieces are now the only reality that seems to matter.

For some years I worked in the USDA South Building, the main headquarters of the U.S. Department of Agriculture in Washington, DC. Built originally to be a federal prison, it is a huge edifice, having, I have been told, more than 13 miles of corridors. I occasionally worked in the nearly deserted building late at night, and walking down the silent, dimly lit halls, I sometimes fancied that I was sharing them with the ghost of W.O. Atwater. One thing is certain; after more than 100 years, his spirit still weighs heavily on U.S. agricultural policy.

The foods he believed should be eaten by the working poor are the very ones the government now favors with the biggest subsidy payments. Chief among them are the commodity crops that are easily processed into refined flour, sugar, high fructose corn syrup, and vegetable oils, all major raw materials for making industrial foods. Although a few subsidies are paid to farmers growing fruits and vegetables, USDA dollars go overwhelmingly to grains such as corn and wheat, oilseed crops, and sugar. Table 7.1 lists the ten agricultural products receiving the largest subsidies from the U.S. Department of Agriculture during the past 25 years. Data are from the Environmental Working Group website (farm.ewg.org).

No fruit or vegetable made this list. If we extended it, apples would come in at number 13. Between 1995 and 2020, apple growers received a little more than $261 million in subsidies, a rather paltry amount when compared with the enormous sums received by the growers of crops such as corn, soybeans, rice, and sorghum.

TABLE 7.1 THE TEN MOST HEAVILY-SUBSIDIZED AGRICULTURAL PRODUCTS

Crop or product	Total subsidies paid (1995–2020)
Corn	$116,570,999,819
Soybeans	$44,940,426,399
Wheat	$44,445,878,970
Rice	$16,841,106,496
Livestock	$12,458,279,636
Sorghum	$8,965,549,426
Dairy	$6,374,052,667
Peanuts	$5,904,163,618
Barley	$3,227,560,234
Sunflowers	$1,364,247,406

I sometimes think about Atwater's untimely death. As discussed previously, he suffered a massive stroke at the age of 60 and then lingered in a state of near-helplessness for three years, cared for by his wife and daughter. At the time of his death in 1907, no one had any inkling of just how important vitamins are to human health. Did Atwater follow the same dietary regimen he recommended for the working poor, avoiding fresh fruits and vegetables because he considered them frivolous, unnecessary luxuries? We now know that, because of the vitamins, minerals, fiber, and potassium that fruits and vegetables contain, eating generous amounts of them can help avoid heart disease, hypertension, and strokes. Did Atwater die early and suffer unnecessarily because of his own nutritional zealotry?

Regrettably, Atwater's dietary biases have now become well entrenched in American agriculture and food production. Refined carbohydrates, fats (mostly vegetable oils), and sugar, much of it in the form of high fructose corn syrup, have become the raw materials for our industrial food system. An endless stream of these commodities are trucked from our nation's farms and transformed into thousands of "foodstuffs" for distribution to the public. In the 1950s a large supermarket might have sold a few thousand food items. By the 1970s this had grown to nearly 12,000, a number that now has doubled, with new items appearing on the shelves every week (Bowers 2000). We can be reasonably sure that thousands of new vegetables, fruits, nuts, or edible animal species have not miraculously appeared on Earth since the 1960s. Instead, these additions consist mostly of industrial foods — new iterations of refined flour, corn meal, sugar, high fructose corn syrup, and chemically-altered vegetable oils.

Interestingly, much of the academic world seemed to have little understanding of the revolutionary changes that were taking place in America's food supply system during the 20th century. In 1946, Oxford University was offered a large sum of money to found an institute of human nutrition. The university turned it down. They believed that after 10 years or so, there would be nothing left to study in nutrition (Hall 1976). The consensus among scientists seemed to be that, when it came to food and nutrition, there was little left to be learned. According to two 20th century authorities,

> Nutrition research has progressed until it is now possible to devise a diet consisting solely of highly purified and many synthetic components... Nutrition has thus far been successful in its search for an answer to the question of what constitutes an adequate diet and how such a diet can be compounded from the foodstuffs available to man and domestic animals. — A.E. Harper, University of Wisconsin, 1969

> [T]he developments of modern science make it immaterial what sort of fat or oil is available; any one sort can be refined, deodorized and, if necessary, hydrogenated so as to convert it into a useful, uniform article. — Magnus Pyke, Research Station Director, Scotland, 1970

Factory foods or "foodstuffs" do have some practical (not nutritional) advantages over real foods. Many real foods spoil more quickly and often have to be washed, peeled, sliced, or cooked for a long time. Most factory foods are much more convenient. You can take them home and store them in your cupboard, confident that they will be edible tomorrow or even weeks from now. When it comes time to eat them, they either do not require cooking or can be prepared very quickly in a microwave oven. Just like fast food restaurants, modern supermarkets feature convenient, high-calorie foods concocted from refined grains, cheap vegetable oils, and high fructose corn syrup — all subsidized by the American taxpayer. Factory foods sold in supermarkets are fast food in another guise, designed for convenience, long shelf life, and ease of preparation. In addition, they are made up largely of three things that humans love: fat, refined carbohydrates, and sugar — or in more recent years, high fructose corn syrup.

Consider some statistics. Americans now spend less than half of each food dollar on food to be prepared and served at home; and only ten percent of that goes for foods in their natural state. Most food purchases at supermarkets consist of packaged, factory-made foodstuffs. Only about five percent or so of the total food dollar goes to buy such things as apples, bananas, raw potatoes, tomatoes, and chickens. Most grocery carts are overflowing with items such as soft drinks, canned spaghetti, cookies, cakes, pastries, frozen pizzas, chips, dry cereals, and canned soups (USDA 2002).

The makers of fast food grocery items compete the same way that fast food restaurants do. They have learned that whoever makes shopping and cooking most convenient and sells foods that are highest in fat, sugar, and salt for the lowest price will win. The list below illustrates the 20th century shift toward high-calorie convenience foods by highlighting some of the more interesting and significant innovations or food introductions in the order they appeared. The list is based on an Economic Research Service article titled "A Taste of the 20th Century" (USDA 2000).

Year	Innovation
1901	A&P food chain, 200 stores
1903	Kellogg's sugared corn flakes
1903	Pepsi Cola
1904	Quaker puffed cereal
1910	Aunt Jemima pancake flour
1912	Oreo cookies
1916	First self-service food store
1928	Peter Pan peanut butter
1928	Velveeta processed cheese
1929	Birdseye frozen foods
1930	Mechanically sliced Wonder Bread
1932	Frito's corn chips
1936	First supermarket shopping cart
1937	Kraft macaroni and cheese dinners
1942	La Choy canned Chinese foods

1954	Swanson frozen TV dinners
1958	Rice-a-roni
1964	Carnation Instant Breakfast
1965	Cool Whip
1967	Countertop microwave oven
1970	Microwave oven
1971	High fructose corn syrup (HFCS)
1974	Wendy's drive through service
1980	HFCS added to Coca Cola
1986	Pop Secret microwave popcorn
1993	SnackWell's cookies and crackers

The 20th century agricultural boom enabled America to grow vast quantities of food; and the modern restaurant and supermarket industries responded by creating thousands of convenient, tasty, calorie-rich food products. The whole country went on an eating binge. This can be verified by examining food disappearance data collected annually by USDA's Economic Research Service. Disappearance data are used to measure the total amounts of different foods that are available to be eaten each year. Let us use a simple example — tomatoes. Each year the USDA records the total amount of tomatoes sold into the commercial market by farmers. They also record the total amount of tomatoes exported. The difference between the total amount sold by farmers and the amount exported "disappears" into the food system and is available to be consumed as food in some form. This information has been gathered for a wide variety of food crops since 1909.

The comparisons between 1950 and 2000 are especially revealing. In 2000, the food system was providing each American with 700 more kilocalories per day than in 1950. After adjusting for wastage or spoilage, Americans were eating nearly 500 more kilocalories each day in 2000 than their parents or grandparents had eaten in 1950. They did this by consuming a lot more meat, a lot more sugar, and a lot more added fat, especially vegetable oils.

From 1950 to 2000, per capita meat consumption went from a little more than 90 pounds each year to about 140. In addition to eating more meat, Americans greatly increased their consumption of added fats and oils, consuming 20 pounds more per capita in 2000 than in 1950. The USDA defines added fats and oils as those we use directly, such as butter on bread, dressings on salads, and the shortening and oils added to commercially prepared cookies, pastries and fried foods. The nation's annual cheese consumption also grew dramatically, from fewer than ten pounds per capita in 1950 to nearly 30 pounds in 2000. Much of this increase can be attributed to the proliferation of pizza restaurants and the popularity of cheese sauces and cheesy snacks.

While increasing their consumption of meat, added fats, and cheese, Americans also indulged their national sweet tooth. In 1950, each of us took in a little over 75 pounds of caloric sweeteners, which are concentrated sugars and syrups derived mostly from sugar cane, sugar beets, and corn. By 2000 this had increased to almost 110 pounds per capita annually. Interestingly, much of this increase took place after 1980, following the introduction of high fructose corn syrup. Americans were consuming almost no high fructose corn syrup prior to 1980, but by the mid-1980s each of us was consuming nearly 20 pounds of this new product each year, mostly in soft drinks and sweetened convenience foods. By 2000, each American was consuming, on average, more than 45 pounds of high fructose corn syrup annually.

By 2000, Americans were taking in 350 more kilocalories each day than they had in 1983, thanks largely to this new product. This was accompanied by a surge in national obesity rates. Soft drinks sweetened with high fructose corn syrup deserve special mention. In 1950 each of us drank about ten gallons of our favorite soft drink; by 2000 this had increased to more than 50 gallons each, most of them sweetened by high fructose corn syrup. Thanks largely but not entirely to soft drinks, the average American now takes in the equivalent of more than 30 teaspoons of sugar each day (Gerrior and Bente 1997; USDA 2002).

One group of industrial foods, which includes such things as cakes, cookies, ice cream, pastries, and candy deserves special mention because they represent a form of food which did not exist on Earth until very recent times. To my knowledge, there is no natural food that contains both large amounts of sugar and large amounts of fat. Fruits and honey are sweet but contain negligible amounts of fat, while nuts and certain animal parts contain a lot of fat, but no sugar. Mother Nature was wise to keep these two apart. Both fat and sugar, when eaten alone, are largely self-limiting.

Have you ever tried to eat a lot of extremely sweet or very fat, greasy food at one sitting? In both cases, you probably stopped after eating only a moderate amount.

But when fat and sugar are mixed together, the appeal and palatability of both are enhanced. Sugar makes fat taste better, while at the same time, the admixture of fat keeps foods containing as much as 70 percent sugar from tasting too sweet. Combining these two created a class of foods for which evolution has not prepared us. One scientist found that the perfect laboratory mixture to make rats extremely obese in a very short time consisted of 40 percent fat, 40 percent sugar, and 20 percent refined flour. Fat alone did not quite do the job; in the words of the researcher, "It took the addition of sugar to light the fuse" (Arnot 1997). Many humans have made themselves obese, sometimes enormously so, by making these beguiling fat-sugar combinations a large part of their diets.

Supersized Nation

Few things are more rewarding than a great idea, and filmmaker Morgan Spurlock came up with a truly inspired one in 2002. While watching television after a huge Thanksgiving dinner, he became intrigued by a news story about two girls who were suing McDonald's because, according to their complaint, eating there had made them fat and unhealthy. In response, a McDonald's representative argued that their food was nutritious. It was at this point, said Spurlock, that "the bells went off." He decided that for one whole month he would eat every meal — breakfast, lunch, and dinner, at a McDonald's restaurant. In addition, if the person behind the counter or the drive-through speaker asked, "Sir, would you like to supersize that?" Spurlock had to say yes.

Before beginning this regimen, Spurlock was checked out by three medical doctors and a workout coach, who all pronounced him to be in excellent health and to be very fit. He began filming his excursions to McDonald's as well as his periodic visits to the medical personnel who monitored his health during the month-long experiment. At the start of his fast-food adventure, Spurlock was six feet, two inches tall and weighed 185 pounds, with cholesterol and triglyceride levels in the ideal range.

At least one of the doctors cautioned him that pursuing this course for a month might cause his triglyceride levels to rise a bit and, by the way, he might gain a few pounds — what a classic failure of imagination! It was so much worse than anyone would have predicted. By the end of the month, Spurlock's weight had ballooned to well over 200 pounds, nearly one pound for every day he was on the fast food diet. In addition, his cholesterol level had risen by 60 points and he had started showing signs of liver deterioration. His doctors became so alarmed that they advised him to quit before the month was up. Instead, he persevered, kept eating fast food and kept filming his rapid descent into dietary self-destruction. The result was a graphic, highly-entertaining documentary with a message that goes well beyond one man's personal experience.

It is interesting to note what happened after Spurlock ended his fast food experiment and returned to his pre-McDonald's diet of natural foods. He quickly returned to the weight at which he had been before the experiment and his cholesterol and metabolic markers returned to normal. Had he continued eating the fast food diet, he eventually would have stabilized at a new set point, with a body weight of 200 to 220 pounds or more. In addition, his health would have continued to decline. In addition to becoming increasingly obese, he most likely would have developed diabetes and hypertension — and in a decade of so, would have been a prime candidate for heart disease.

In addition to its other merits, Spurlock's 2004 film can serve as a revealing metaphor for America as a whole. During the 20th century, and especially during the latter half of it, our entire nation was supersized. America is growing obese and diabetic at an alarming rate. Obesity rates in America increased only moderately from 1960 to 1980, but then things changed drastically. From 1980 to 2020, adult obesity rates surged from a little over 15 percent to over 40 percent. Overall, more than two in three American adults now are classified as overweight. Four out of ten are obese, while nearly one in ten is morbidly obese.

These numbers continue to rise. The change among children is most alarming. In 1980 about five percent of American children were obese. The rate is now approaching 20 percent. If such trends continue, nearly every American will be overweight by the year 2050, and more than half of us will be clinically obese. As many as two Americans in ten will weigh more than 500 pounds, and people tipping the scales at half a ton will not be uncommon. At least one-third of us, and probably more, will be undergoing treatment for diabetes (Kreuger 2005; Simmons 2007).

The Three Ages of Food

Eaton and Konner (1985) estimated that, on average, hunter-gatherers got about 65 percent of their calories from plant foods and 35 percent from animal foods. They also estimated that the hunter-gatherer diet consisted of about 45 percent carbohydrates, 20 percent fat, and 35 percent protein. Hunter-gatherers ate a wide variety of foods throughout the year. Their diet commonly included wild meat, vegetables, insects, shellfish, fruit, nuts, bird eggs, and any other tasty item they might encounter as they traversed their territory. Overall, they consumed about four to six pounds of natural foods each day, depending on their body size and level of activity.

Adult humans ate this approximate weight of food each day for tens of thousands of years. This brings up a question some might consider simplistic; why four to six pounds of food? Why not two pounds or, at the other extreme, why not ten or twelve pounds? It all comes back to the energy density of natural foods and the energy requirements of our bodies.

We need energy from food to keep our systems functioning and to fuel our muscles so we can move about and do work. The energy required to keep the body running even if we are not active — for example, when we are sleeping — is referred to as basal metabolism. All of the body's organs and systems need a certain amount of fuel just to keep the life processes going. A typical man weighing 150 pounds needs about 1,800 kilocalories each day just to meet his basal metabolic needs, while a typical woman weighing 120 pounds needs about 1,300 kilocalories.

But most people do not stay in bed all of the time. In addition to just keeping their body functioning, they burn additional energy moving about and doing work. The total energy expenditure of active adult humans typically ranges from 1.5 to 2.0 times their basal metabolic rate. Thus a moderately active man weighing 150 pounds needs a total of about 2,700 kilocalories each day (1,800 times 1.5). Actually, he would need closer to 3,000 kilocalories, since the body is not totally efficient at converting food to energy (Harris and Benedict 1918).

Assume that the daily energy requirement of an adult hunter-gatherer man was 3,000 kilocalories — not an unreasonable assumption. The question then becomes: How much food (that is, what weight or volume of food) would he have needed to eat each day? Fortunately, this is not hard to figure out. Most of the natural foods eaten by early humans are still around, and modern science allows us to measure their energy content with precision.

Table 4.1 in Chapter 4 listed some foods known to have been eaten by hunter-gatherers and the amount of food energy provided by a pound of each. On average, the mix of wild game, fish, fruits, vegetables, and nuts eaten by hunter-gatherers yielded about 700 to 800 kilocalories per pound. A simple calculation shows that a hunter-gatherer could take in 3,000 kilocalories by eating four to five pounds of such foods, since 3,000 kilocalories divided by 700 kilocalories per pound (3,000/700) equals 4.28 pounds of food. The four to five pounds of food eaten daily by our 150-pound hunter-gatherer represent about two to three percent of his body weight.

Our species lived as hunter-gatherers for more than 7,000 generations. During that time the average energy density of the human diet did not change appreciably. As humans moved into new regions, they encountered new plants and animals; but the meat of a deer in central Europe differed little in calorie content from that of a gazelle in the Middle East or a buffalo in North America. The same was true of vegetables and fruits, which were pretty much the same the world over. Interestingly, the amount of natural foods eaten by humans (two to three percent of body weight) is also the norm for several other species, including cats, dogs and cattle.

Human metabolism was shaped and fine-tuned by this long existence as hunter-gatherers. We evolved to have a body of a certain mass and size, fueled by a relatively constant intake of food each day — food which, in the aggregate, had about the same energy density for most of our time on earth. But the makeup of the human diet and its energy density changed drastically around 10,000 years ago with the adoption of agriculture. The change was especially drastic following the rise of feudalism, which began a few thousand years later.

Feudalism brought about a revolutionary change in the quality of the human diet. Remembering that stone-age hunters and gatherers got about 600 to 700 kilocalories from each pound of food they consumed, let us consider what happened in peasant economies. When farming and peasant life became fully established, most people on earth shifted from a mixed diet of wild plants and animals to one consisting almost exclusively of cooked grains — corn, rice, or wheat for the most part. There were some exceptions; for example, potatoes were the staple food of many peasants in the Andean region of the Americas.

Table 5.1 in Chapter 5 listed some typical foods eaten by peasants and the amount of food energy provided by a pound of each. The data in Table 5.1 show that peasant diets provided about 450 to 550 kilocalories per pound, significantly less than the 600 to 700 kilocalories per pound provided by the hunter-gatherer diet.

As evidenced by this list, the caloric density of human foods actually declined after the adoption of agriculture. On average, the mostly grain based diets eaten by peasants were much less calorie dense and much less nutritious than the diets enjoyed by hunter-gatherers. The peasant diet was notably deficient in fat and protein. For example, peasants living in the rice-growing regions of Asia and the maize-growing areas of Mexico and Central America consumed less than 15 percent (some as low as ten percent) of their daily calories in the form of fat. The diet of Japanese peasants contained around ten percent fat, while Chinese peasants consumed about 15 percent of their calories in the form of fat. In contrast, fat accounted for more than 20 percent of the calories in most hunter-gatherer diets. In addition to being lower in fat, the typical peasant diet contained very little high quality animal protein.

With the transition from hunting and gathering to peasant life, most humans were forced to consume a diet that was much less energy dense than the one with which our species evolved. To use an automotive analogy, our fuel mix became leaner. The recent change to high-calorie factory foods was just the opposite; the human race began consuming a much richer fuel mix than the one under which we evolved.

During the last half of the 20th century, former peasants and their descendants began eating an energy-rich diet of factory-produced foods high in animal protein, fats, sugar, and refined carbohydrates. In only a few generations, much of the human race nearly doubled the energy density and protein content of its food supply. Americans have led the way, with many of us now taking in more than 1,000 kilocalories in every pound of food we eat; some of us are consuming a diet that is even more energy dense — as high as 1,500 kilocalories per pound, nearly three times the Paleolithic and peasant average. Table 7.2 shows the number of kilocalories in a pound of some representative industrial foods, the kinds eaten by millions of Americans at nearly every meal. Data are from the USDA nutrient database, FoodData Central (fdc.nal.usda.gov).

If you compare the modern American diet with that eaten by hunters-gatherers or peasants, several things stand out. First, we are eating nearly two times as much fat as hunter-gatherers did and three to four times as much fat as peasants. Second, we are eating an astounding amount of refined sugars. The average American gets nearly 700 kilocalories daily from sugar and high fructose corn syrup alone. In contrast, hunter-gatherers and early peasants ate no refined sugar; peasants living in more recent centuries might have had some access to it, but very little. Large amounts of sugar and fat make the modern diet very energy dense. Our species has gone from a hunter-gatherer diet providing 600 to 700 kilocalories per pound of food to a peasant diet yielding only 450 to 550 kilocalories per pound; and more recently to an industrial diet that typically yields 1,000 to 1,500 kilocalories or more per pound.

Table 7.2. Kilocalories in one pound of common industrial foods

Food	Kilocalories in one pound
Cherry pie	1,435
Chicken nuggets	1,710
Double cheeseburger	1,280
Dry cereal	1,670
French fries	1,280
Glazed donuts	1,890
Oatmeal cookies	2,002
Peanut butter	2,760
Pepperoni pizza	1,350
Potato chips	2,415
Processed meats	1,360
Salad dressing	1,900
Candy bar	2,250

Many scientists now believe that, in addition to increasing the caloric density of the modern diet, sugar is directly linked to the increased prevalence of diseases such as diabetes, hypertension, heart disease, and cancer. The next chapter describes how consuming large amounts of sugar can disrupt the metabolic and hormonal balance of the body, leading to the so-called "diseases of civilization."

The next chapter also considers the relative amounts of potassium (K) and sodium (Na) in our diet. Modern humans are the only free-living species of land animal that consumes more sodium than potassium. For most of human evolution, our species consumed five to ten times more potassium than sodium. Modern Americans, because so much salt is added to commercial foods, now consume two to three times more sodium than potassium. Many scientists believe that this recent reversal is playing havoc with our sodium potassium pumps, which generate the electric current needed to keep our bodies working properly. Too much sugar, too much fat, too much sodium, too many calories, and not enough potassium — all of these, taken together, seem to be killing us. The next chapter explains how.

Chapter 8. Dead Presidents

American presidents, while performing extremely stressful jobs, are coddled in many ways, with ready access to the best food and the best health care. Also, because they are such public figures, we know exactly when and at what age they died, and have a pretty good idea of what killed them — often more than we would care to know, as shown by the following excerpt from the obituary of president Ulysses S. Grant, published in the July 24, 1885 issue of the *New York Times*.

> He coughed somewhat after midnight, and was able with the doctor's aid to dislodge the mucus and throw it off, but from about 3 o'clock he could neither dislodge it or expectorate, and it began to clog his throat and settle back into his lungs. It was about 4 o'clock when the rattle in the throat began. Dr. Shrady...had been giving hypodermics of brandy with great frequency, and applying hot cloths and mustard to various parts of the body...It was soon evident that the General was too far gone to be aided by stimulants. Then came the waiting for death.

Interestingly, Ulysses S. Grant is the only American president to have died of cancer, although three other presidents have been treated for it. The age at which our presidents died and the manner of their deaths tell us a lot about the nature of life and death in America. Table 8.1 shows the cause of death for each American president who died of natural causes during the 20th century before reaching the age of 70.

TABLE 8.1. CAUSE OF DEATH FOR PRESIDENTS WHO DIED DURING THE 20TH CENTURY BEFORE AGE 70

President	Year of death	Cause of death
Benjamin Harrison	1901	Pneumonia
Theodore Roosevelt	1919	Heart attack
Warren G. Harding	1923	Heart attack
Woodrow Wilson	1924	Cardiovascular disease
William Howard Taft	1930	Cardiovascular disease
Calvin Coolidge	1933	Heart attack
Franklin D. Roosevelt	1945	Stroke
Lyndon B. Johnson	1973	Heart attack

Benjamin Harrison died of pneumonia in 1901. All of the remaining seven are reported to have died from one or more complications of cardiovascular disease, such as a heart attack or stroke. Heart disease also plagued other 20th century presidents. Dwight Eisenhower survived a major heart attack in 1955. More recently, surgeons performed heart bypass surgery on former president Clinton, while former president George W. Bush had a stent installed to correct arterial blockage.

Diseases of Civilization

Presidents who died of so-called natural causes during the 20th century are an accurate reflection of what was taking place in the rest of America as well. While cardiovascular disease was killing nearly all of them, it was doing the same to many of their fellow citizens. By 1950, cardiovascular disease had become the leading cause of death in the nation. On the way to dying of heart disease, 20th century Americans increasingly were plagued by accompanying ailments that had been relatively rare before 1900, such as obesity, diabetes, and hypertension, which are commonly referred to as "diseases of civilization." Such ailments were rare prior to the 20th century, becoming common only after a large proportion of the human race moved into cities and began eating industrially-produced foods.

In 1900, more than half of American deaths were the result of infectious diseases, mainly tuberculosis, pneumonia/influenza, and diphtheria. However, the so-called diseases of civilization, such as heart disease, cancer, and cerebrovascular disease, were becoming significant. Table 8.2 shows the ten most common causes of American deaths in 1900, according to the Centers for Disease Control website.

TABLE 8.2. THE TEN MOST COMMON CAUSES OF DEATH IN THE UNITED STATES IN 1900

Pneumonia/influenza
Tuberculosis
Gastrointestinal disease
Heart disease
Cerebrovascular disease
Nephropathies
Accidents
Cancer
Senility
Diphtheria

In 1900, infectious diseases such as pneumonia, tuberculosis, and diphtheria were killing twice as many people as metabolic ailments such as heart disease, cardiovascular disease, or cancer. But 100 years can make a big difference. By the beginning of the 21st century, most deaths in the United States were resulting from lifestyle or metabolic diseases. Since such illnesses now are so common and tend to occur together in the same individual, doctors have created a name for them — *cardiometabolic* diseases. Cardiometabolic diseases refer to a cluster of conditions, including hypertension, high blood sugar, abdominal obesity, and elevated triglycerides. The nearly 50 million people in the United States who suffer from cardiometabolic diseases are two times more likely to die from coronary disease and three times more likely to have a heart attack or stroke. Table 8.3 shows the ten most common causes of death in America as of 2010. Note that only one infectious disease, pneumonia/influenza, remains on the list, and it is very near the bottom.

TABLE 8.3. THE TEN MOST COMMON CAUSES OF DEATH IN THE UNITED STATES IN 2010

Heart disease
Cancer
Chronic lower respiratory disease
Stroke (cerebrovascular disease)
Accidents
Alzheimer's disease
Diabetes
Nephropathies
Pneumonia/influenza
Suicide

According to the Centers for Disease Control, most children born in 21st century America will develop one or more cardiometabolic diseases during their lives. On the way to dying of such things as heart attack, stroke, kidney failure, or cancer, two-thirds of them will be overweight or obese, and many will be diabetic; increasing numbers of children will become obese as toddlers or in elementary school. More than 30 percent of boys and nearly 40 percent of girls born in the 21st century will become diabetic during their lifetimes. For Black and Hispanic children, the prognosis is even worse; more than half of them will suffer from diabetes (Narayan et al. 2003).

According to the Centers for Disease Control, a man who is diabetic at age 40 will have his life shortened by more than ten years. A woman diagnosed with diabetes at 40 can expect to die 15 years earlier than her non-diabetic peers. People who are obese and diabetic are far more likely to die of a heart attack, stroke, cancer, or kidney disease at an early age. As obesity and diabetes affect more and more people at earlier and earlier ages, this will impact the overall life expectancy of our nation. A recent report concluded that if something is not done, life expectancy at birth in America could decline by as much as five years in only a few decades (Olshansky et al. 2005). For detailed statistics and other information on cardiometabolic diseases, see Benjamin et al. (2019), Dwyer-Lindgren et al. (2016), and Hales et al. (2018).

There are two competing theories as to why cardiometabolic diseases began to reach epidemic levels in the 20th century. One theory — the one favored by most of the medical establishment, is that dietary fat, especially animal fat, is the problem. The Economics Research Service (USDA) has been tracking the amount of fat, protein, carbohydrate, and other nutrients in the American food supply since 1909. Measuring the amount of each nutrient actually consumed would be nearly impossible. Instead, USDA determines the amounts of several hundred foods that are available for consumption on an annual basis. They do this by keeping records of commodity flows from production to final end uses. By subtracting exports, industrial uses, and end-of-year inventories, they are able to calculate the per capita amount of each food that the economy makes available for human consumption during a given year. Such estimates are commonly referred to as food disappearance data. Using food composition tables, they then can estimate the per capita amounts of specific nutrients, such as fat, protein, or carbohydrates that are available for consumption during a given year.

In 1909, the food industry was making about 100 pounds of dietary fat available to each American during the course of a year. By 1994, this had increased to about 130 pounds per person. During the same 85-year period, per capita availability of dietary protein increased from 80 to 88 pounds per year. The amount of dietary carbohydrate that made its way into the economy each year declined slightly — from about 400 pounds per capita in 1909 to 395 pounds in 1994. During the 20th century, Americans increased their consumption of dietary fat by about 30 percent, protein consumption by about 10 percent, and overall carbohydrate consumption not at all. Available food energy increased from 3500 kilocalories per day in 1909 to 3800 kilocalories in 1994.

W.O. Atwater, in his early calorimetric studies at the United States Department of Agriculture, determined that fat provides about 9 kilocalories of food energy per gram, while carbohydrate and protein provide only about 4 kilocalories. The body also is very efficient at storing fat. According to Dean Ornish (1993), "One hundred fat calories can be stored as body fat by expending only 2.5 calories, whereas your body must spend 23 calories — almost ten times as much — to convert 100 calories of dietary protein or carbohydrates into fat." Based on thermodynamics alone, it seemed obvious that those who ate a lot of fat would take in more calories and be more likely to become overweight or obese. Since it is well known that obese people are likely to be diabetic, many scientists assumed that eating excessive amounts of fat was the primary cause of diabetes. It also was determined that fat, especially animal fat, caused blood cholesterol levels to rise. Since cholesterol deposits are found on the walls of diseased arteries, eating too much fat, especially animal fat, was assumed to be the primary cause of heart and arterial disease as well.

Perhaps the best known scientist to focus on dietary fat as the cause of cardiometabolic disease was Ancel Keys. Chapter 3 describes the starvation experiments he carried during World War II. Shortly after the war, Keys focused his attention on heart disease, and by the early 1950s was convinced that he had discovered its cause. In 1955, he presented his "diet-heart hypothesis" to a meeting of the World Health Organization in Geneva, Switzerland. As part of his presentation, he displayed a graph showing the relationship between deaths from heart disease in 1948-1949 and the percentage of fat in the diets of six countries — Japan, Italy, England and Wales, Australia, Canada, and the United States. Fat accounted for about 10 percent of calories in the Japanese diet, and only about one Japanese in 1,000 was dying of heart disease during a given year. The English were eating three times as much fat as the Japanese, and their rate of death from heart disease was nearly four times greater. Americans, at the very top of Keys' graph, were eating four times as much fat as the Japanese, and were nearly eight times as likely to die of heart disease (Keys 1956).

Some, however, were not convinced by these numbers, arguing that Keys appeared to have cherry-picked his data, selecting only those countries that fit his hypothesis. In addition, they pointed to other factors that showed as strong a correlation with heart disease mortality as did fat consumption. These included the amount of sugar or protein consumed and the number of cars or television sets sold per capita. In fact, almost any factor that was correlated with post-World War II prosperity would show a strong statistical correlation with heart disease (Yerushalmy and Hilleboe 1957).

In response to his critics, Keys launched the Seven Countries Study, which tracked the rates of heart attack and stroke in more than 12,000 middle-aged men in seven countries. Early results seemed to support the diet-heart hypothesis. Men eating the most animal fat had the most cholesterol in their blood, and those with the most cholesterol in their blood had the highest rates of coronary heart disease (Keys et al. 1980). In Keys' mind, the question had been settled; eating fat, especially saturated animal fat, causes cholesterol levels to rise, and rising cholesterol levels damage arteries, causing coronary disease.

People who died of heart disease almost always had cholesterol deposits in the plaque on the walls of their arteries. Because of this association, cholesterol had long been suspected of playing a role in heart and arterial disease. Jeremiah Stamler, a highly-regarded authority of the time, likened cholesterol to "biological rust" that could "spread to choke off the flow [of blood] or slow it just like the rust inside a water pipe" (Blakeslee and Stamler 1966).

Early animal experiments supported this view. In 1913, Russian scientist Nikolaj Anitschkow caused atherosclerotic lesions to form in the arteries of rabbits by feeding them large amounts of cholesterol. Other scientists repeated this experiment in other animals, and by 1950 it was widely assumed that eating high cholesterol foods such as eggs would lead to cholesterol buildup and arterial disease. But this idea soon was proven to be wrong. Ancel Keys (1952) found that, even when he fed large amounts of cholesterol to experimental subjects, the levels of cholesterol in their blood remained nearly constant. A Swedish doctor conducted an experiment on himself by eating eight eggs each day for nearly a week. Surprisingly, he found that the amount of cholesterol in his blood, instead of going up, actually declined slightly (Ravnskov 2000).

Scientists are not sure why dietary cholesterol has so little effect on the amount of cholesterol in the blood. One theory is that, if you consume a lot of cholesterol, your body simply slows down its own production to compensate. Some have questioned the use of rabbits and other herbivores in such studies because they do not eat things such as eggs or meat in nature. One should not expect their bodies to have evolved the capacity to process cholesterol and clear it from the body.

The invention of gas-liquid chromatography in the early 1950s made it easier to separate different kinds of fatty acids and to assess their effects on the human body. In a very short time, scientists working in California, Massachusetts, and in Europe found that replacing animal fats in the diet with vegetable oils would quickly lower blood cholesterol levels. Ancel Keys took note of such studies and, in the words of science journalist Nina Teicholz (2014), "confidently drew a line of causation from saturated fat in the diet to serum cholesterol in the blood to heart disease." This was the core of his diet-heart hypothesis — the idea that eating fat, especially saturated fat, caused cholesterol levels to rise, and that increasing amounts of cholesterol in the blood resulted in artery damage and heart disease. Keys proclaimed at a meeting in 1954,

> "No other variable in the mode of life besides the fat calories in the diet is known which shows anything like such a consistent relationship to the mortality rate from coronary or degenerative heart disease." (Keys and Anderson 1954).

Keys' career and his diet-heart hypothesis were given a big boost about a year later when president Dwight Eisenhower suffered the first of several heart attacks. Eisenhower's personal physician was Paul Dudley White, a professor of medicine at Harvard and one of the founders of the American Heart Association. White had written a textbook on heart disease in 1931 and had worked closely with President Harry Truman to establish the National Heart Institute. Needless to say, he was a very powerful figure in American medicine.

Keys had a penchant for cultivating influential people to advance his career. For instance, he parlayed his research for the military during World War II into a four-year appointment as a special assistant to the Secretary of Defense. Paul Dudley White had accompanied Keys and his wife on several overseas trips to gather fat and cholesterol data in support of the diet-heart hypothesis. By the time Eisenhower had his first heart attack in 1955, Keys appears to have persuaded White that the best way to ensure cardiac health was to eat a diet very low in fat. One of Keys' former associates stated in an interview many years later (Teicholz 2014), "You have to understand what a persuasive person Keys was. He could talk to you for an hour and you would utterly believe everything he said."

White put Eisenhower on the low-fat, low-cholesterol diet advocated by Keys, and in his frequent reports to the public on the president's condition, advised other Americans to follow the same regimen. In a New York Times article, White cited Keys' dietary theories and described his work as "brilliant." The publicity surrounding Eisenhower's heart attack and subsequent treatment convinced many people that, in order to prevent heart disease, they should eat as little fat as possible and that they should replace animal fats with vegetable oils.

In 1961, the American Heart Association came out in favor of Keys' diet-heart hypothesis, recommending that everyone adopt the so-called "prudent" diet, one low in cholesterol and saturated animal fats. Soon after that, Keys was featured on the cover of *Time Magazine*, wearing a white lab coat and with his image superimposed on a realistic drawing of the human heart. The accompanying article referred to Keys as "Mr. Cholesterol." Nine years later, largely due to Keys' influence, the American Heart Association recommended that, in addition to saturated fat and cholesterol, the total amount of fat in the diet should be restricted.

Ancestral Diets

For many centuries prior to the 20th century, most people on Earth were eating the kind of diet recommended by Keys and his supporters, a grain-based diet high in carbohydrates and low in fat. Humans began to transition from hunting and gathering to agriculture about 10,000 years ago, and by 5,000 years ago, economies and cultures specific to certain grain crops had developed in various parts of the world. Wheat became the staple food in the Middle East, while the agriculture of Mexico and Central America was based largely on corn. Chinese agriculture began with millet production on deep, loess soils so common in parts of China, followed a few thousand years later by irrigated rice farming along major rivers. Populations grew rapidly under agriculture, and such large numbers could only be fed with a grain based, high-carbohydrate diet.

At the beginning of the 20th century, most of the world's people remained poor and even in industrialized countries, ate high-carbohydrate diets consisting mostly of wheat, corn, rice, potatoes, and beans, with very little fat or animal protein. On a worldwide basis, agricultural production barely met human needs, so nearly all of the grain produced had to be consumed directly by humans, with only about ten percent fed to cattle, hogs or other meat producing animals. As late as 1910, bread and potatoes still accounted for nearly half of the calories consumed by urban dwellers in both the United States and England. On the eve of World War II, the poor of England still ate mostly bread, with each person consuming about a pound and a half daily (Mount 1975).

People in a number of East Asian countries continue to eat high-carbohydrate diets, yet have very low rates of obesity. Most of these carbohydrates come in the form of rice, which remains a staple food throughout much of Asia. In Taiwan, chopsticks standing upright in a mound of rice symbolize the death of a loved one. In Singapore, a secure job is referred to as an "iron rice bowl," while losing one's job is described as having a "broken rice bowl."

Perhaps no country epitomizes the Asian connection to rice more than Japan. The Japanese word *gohan* can refer just to rice, to any meal, or to food in general, just as we in the West refer to "our daily bread." Breakfast in Japan is *asa gohan*, or simply morning rice. Even the names of Japan's largest companies evoke a past of peasants and rice fields. The Japanese character for Toyota, originally *Toyoda*, means "bountiful rice field," while the character for Honda, the other auto giant, means "main rice field" (Williams 1996).

The fact that the names of two of Japan's most successful companies have a direct connection to rice illustrates the importance that this grain has played and continues to play in Japanese life. According to Keys' Seven Countries Study, the rice-based Japanese diet in 1948–1949 was providing more than 70 percent of the nation's calories in the form of carbohydrates. This level of carbohydrate consumption was somewhat higher than the historical norm because Japan still was recovering from the ravages of World War II; but in the traditional Japanese diet, going back hundreds of years, carbohydrates have provided more than 60 percent of food energy — and most of those carbohydrates have come from white rice.

Despite eating a high-carbohydrate diet for many centuries, obesity has always been rare among the Japanese, and remains so even today. According to the World Health Organization (WHO), only about four percent of Japanese were obese in 2018, compared with more than 40 percent of Americans. In addition to experiencing low rates of obesity, the Japanese have the second highest life expectancy in the world, while the United States comes in at number 42. In addition to Japan, one could cite a number of other Asian countries where people eat high-carbohydrate diets, yet have low rates of obesity; these include South Korea, Cambodia, Okinawa, Thailand, and Vietnam — which, incidentally, has the lowest rate of obesity in the world at around two percent.

It is clear that, for hundreds of years, humans around the world remained lean and healthy throughout life while eating high-carbohydrate, low-fat diets based on such foods as wheat, corn, rice, potatoes, and legumes. It also is clear that, during the 20th century, Americans began to eat increasing amount of fat relative to carbohydrate and protein. At the same time, Americans began to consume larger amounts of animal foods such as meat and cheese. In 1950, the typical American was eating about 135 pounds of meat per year; by 1999, this had increased to nearly 200 pounds. At the same time, we more than tripled our consumption of cheese, going from an annual consumption of about eight pounds per person in 1950 to more than 28 pounds per person in 1999 (USDA 2000).

The case seems pretty strong. Fat contains twice as many calories as carbohydrates or protein and the body can store fat at very little metabolic cost, so it follows that the more fat you eat, the more obese you will become. Obesity, in turn, is strongly linked to diabetes. In addition to making us obese and diabetic, eating animal fat raises blood cholesterol levels, making fat a prime suspect in heart disease as well. If you visit the websites of almost any mainstream health organization, such as the American Heart Association or the Centers for Disease Control, you will be advised to limit the amount of total fat you eat and, whenever possible, to replace animal fats with vegetable oils.

The Hormonal Theory

Despite the near unanimous endorsement of grain-based, high-carbohydrate diets by the medical establishment, a number of reputable researchers and physicians are not convinced, arguing that when it comes to cardiometabolic disease, fat has been falsely accused and falsely convicted. They have a different theory and they make an equally compelling case. In contradiction to those who say that everyone must eat a plant-based, low-fat diet to remain healthy, they point to a number of traditional societies that consumed high-fat diets made up almost entirely of animal foods for many generations. Despite eating a diet high in fat and containing almost no carbohydrates, they remained in robust health throughout their lives. The two best known examples are the Inuit people of the Arctic, who live by fishing and hunting marine mammals, and the Maasai cattle herders of East Africa.

In 1906, anthropologist Vilhjalmur Stefansson lived for an entire year with the Mackenzie River Inuit in the Canadian Arctic. His hosts taught Stefansson how to hunt and fish, and he followed their traditional diet for an entire year, eating almost nothing but meat and fish. For more than six months he ate mostly caribou, followed by several months of eating almost nothing but salmon. For nearly a month in the spring, he ate only eggs (Stefansson 1921). It has been estimated that more than 70 percent of the calories in Stefansson's diet came from fat. The Inuit with whom he lived had eaten this same high-fat, high-cholesterol diet all of their lives. In addition, they spent several months each year in a largely sedentary state, with "no real work to do," unable to hunt or fish because of the total darkness. As Stefansson noted (1946), "They should have been in a wretched state... But, to the contrary, they seemed to me the healthiest people I had ever lived with."

Several years after returning from the Arctic, Stefansson and a colleague replicated his experience by eating a diet of nothing but meat, fat, and water for an entire year while under medical supervision. Both men completed the regimen and, upon its completion, proclaimed that they felt extremely well. Physicians who supervised the experiment reported that, at the end of the year, they could find nothing medically wrong with either of the men (Lieb 1929). Stefansson remained on an exclusive diet of animal foods for the rest of his life, was rarely ill, and lived into his 80s.

Kenya and northern Tanzania are home to a pastoral tribal group called the *Maasai*. The original homeland of the tribe is believed to have been the Nile Valley. The Maasai began moving south in the 15th century to find new grazing land for their expanding cattle herds. They conquered indigenous tribes, took their land and cattle, and soon spread throughout the Rift Valley region. By the 1700s they had reached what is now Kenya and Tanzania. Many of the Maasai maintained their traditional nomadic, cattle-herding life well into the 20th century.

Gerald Shaper, a South African physician, began studying the diet and health of a group of Massai called the *Samburu* around 1960. He found that young Samburu men typically drank two to five liters or more of milk each day, depending on the season. This was equivalent to a pound or more of butterfat daily. At times, the young men would add two pounds or more of meat to their daily intake, raising their fat and cholesterol consumption even higher. Despite this, Shaper (1962) found the men to be very healthy, with little evidence of heart disease.

A short time after Shaper began conducting his studies, a medical team led by George Mann of Vanderbilt University took a mobile laboratory to a different part of Kenya to study another group of Maasai. Like their fellow tribesmen to the north, these Massai drank up to five liters of milk a day and considered fruits and vegetables fit only for cattle food. When the milk supply ran low during dry seasons, they would mix it with cattle blood. They also ate lamb, goat, and beef on a regular basis. Fat provided more than 60 percent of calories, and all of it came from animals, meaning that much of it was saturated. Among the young men of the warrior or *murran* class, "no vegetable products were taken" (Mann et al. 1964).

Both Shaper and Mann found the Massai warriors to be very healthy, despite eating a diet that most American physicians believed would condemn them to heart disease and an early death. If saturated animal fats were as deadly as Keys and many others believed, there should have been an epidemic of heart disease among the Massai, but both Mann and Shaper found just the opposite. Body weight and blood pressure of the Samburu tribesman examined by Shaper were 50 percent lower than comparable American men, and unlike Americans, the blood pressure of the Samburu did not rise with age. Mann performed electrocardiograms on 400 men and found no evidence of heart disease. Shaper performed the same test on 100 Samburu men and found only two instances of "possible" heart disease. Mann conducted autopsies on 50 Massai and found only one "unequivocal" example of artery obstruction (Mann et al. 1972).

Ancel Keys was dismissive of these observations, arguing that there really was nothing to be learned from the Inuit of Canada or the Masaii of Africa. Concerning the Inuit, Keys (1970) stated that, although "their bizarre manner of life excites the imagination," especially the "popular picture of the Eskimo...happily gorging on blubber," they "did not demonstrate an exception to the diet-fat coronary heart disease hypothesis." Similarly, he dismissed the work of Mann and Shaper in Kenya by asserting that the "peculiarities of these primitive nomads" had no relevance to understanding heart disease in modern societies.

Keys' offhand dismissal of studies that contradicted his viewpoint was both unscientific and unprincipled, but it was all too typical of Keys. Throughout his career, anyone who challenged his ideas was subject to vigorous, often personal attacks. Shaper and Mann both were very competent scientists, and their observations were very relevant to understanding heart disease and other cardiometabolic ailments. In addition, their conclusions were not new. The idea that people could remain healthy while eating high-fat, animal-based diets has been around for a surprisingly long time.

For more than 150 years, books and scientific articles have argued that is it not eating too much fat that makes us obese and unhealthy; instead, it is the excessive consumption of high-carbohydrate foods such as rice, bread, and potatoes. One of the earliest people to express this view was William Banting, who published his *Letter on Corpulence: Addressed to the Public* in 1863. Banting described how, at the age of 66 and only 5 feet, 5 inches in height, he weighed more than 200 pounds. His hearing and eyesight were failing, and he suffered from an umbilical hernia, chronic indigestion, and other conditions.

Banting consulted noted surgeon William Harvey, who suggested that he severely limit the amount of carbohydrates he ate as a means of losing weight. Harvey apparently based his prescription on the rather simplistic idea that, since grains are used to fatten cattle, eating them might make humans obese as well. Banting followed this advice, proceeding to eat mostly meat, fish, and game, avoiding high-carbohydrate foods such as bread, rice and potatoes. Within a year, he had lost nearly 50 pounds and reported that nearly all of his physical ailments had gone away. In the 4th edition of his pamphlet, published in 1869, Banting described his overall health as "extraordinary," proclaiming, "Indeed, I meet with few men at seventy-two years of age who have so little cause to complain." He continued his low-carbohydrate regimen until his death at age 81, avoiding obesity and living much longer than the average Englishman of his time.

A half century later, in 1919, a New York City physician named Blake Donaldson had become frustrated at being unable to help his obese patients lose weight. The only advice he had to offer them — to cut back on calories, was not working at all. He was inspired to try something new when anthropologists at the American Museum of Natural History in Manhattan told him about the Inuit and how they remained healthy and lean while eating almost nothing but "the fattest meat they could kill."

Donaldson decided that what worked for the Inuit might just work for his patients, so he began putting them on high-fat diets and telling them to stop consuming all foods containing sugar and flour, and instead to eat three meals daily consisting almost entirely of fatty meat. Over the next 40 years, Donaldson prescribed this high-fat, low-carbohydrate regimen to about 17,000 obese patients and, as detailed in his 1961 memoir, *Strong Medicine*, nearly all of them did very well on it, typically losing two to three pounds per week while experiencing little or no hunger. As Donaldson noted in his memoir, he was most impressed by the fact that few of his patients regained the weight they had lost.

In 1944, Donaldson gave a talk at a New York hospital, and Alfred Pennington, the in-house physician for the DuPont Company of Delaware, was in the audience. Pennington was intrigued by the talk and, needing to lose weight himself, tried the carbohydrate-restricted diet recommended by Donaldson. Pleased with the results, he then conducted a study involving 20 DuPont executives. Many of the company's executives were overweight and suffering from heart disease, and Pennington believed that losing weight was the first step in correcting this problem.

He put his test subjects on a diet which included a little over a pound of meat, about half a pound of added fat, and about 60 grams (about 1/10th of a pound) of carbohydrate daily. Most of the men ate about 3,000 calories daily, of which a little over 10 percent came from carbohydrates. The executives lost an average of 7 to 10 pounds per month, while experiencing, according to Pennington (1953), "a lack of hunger between meals" and "increased physical energy and sense of well-being."

Intrigued by this success, Pennington delved into the scientific literature to see if he could find out exactly why the diet worked. His search led him to a group of German and Austrian researchers who, in the 1920s and 1930s, had formulated the "hormonal" theory of obesity. They had observed that men and women store fat differently. Men tend to put on fat around their midsection, while women, at least prior to menopause, tend to store fat below the waist. They also observed that adolescent boys lose fat and gain muscle mass as they go through puberty, while adolescent girls increase their body fat. The body contours of a ten-year-old boy and a ten-year-old girl are very similar, but they differ dramatically by the time both are 15. The German and Austrian researchers reasoned that this could not happen unless fat storage was under some kind of hormonal control.

This viewpoint was given a boost in 1921 with the discovery of the hormone insulin, which soon was recognized as being uniquely fattening. Within only a few years, physicians were treating severely underweight children by putting them on a high-carbohydrate diet and injecting them with insulin. Many of them were able to gain five pounds or more a week. On the other hand, an animal whose pancreas had been removed to shut down insulin production could not gain weight regardless of how much it ate.

According to the hormonal theory, eating a diet high in carbohydrates, especially if they are refined, accelerates the production of insulin. Starch and sugar are broken down into glucose molecules, and the entry of glucose into the bloodstream signals the pancreas to start producing insulin. The higher the glucose load, the greater the insulin production. Insulin induces the body to store fat instead of using it, while preventing previously stored fat from being mobilized. As a result, the body uses carbohydrates for fuel, while stashing away fat and making it unavailable. Eating a lot of refined carbohydrates throughout the day causes insulin levels to remain elevated. Even if blood sugar drops temporarily, high insulin levels keep stored fat from being mobilized. This can result in constant hunger, and trigger cravings for even more carbohydrates.

But eating additional carbohydrates causes insulin levels to spike even higher, leading to even more fat storage followed by more hunger. A person eating a diet high in refined carbohydrates supposedly becomes like a drug addict; the more he or she eats, the more the body craves, creating an endless cycle of dependency, or "carbohydrate addiction," which leads to even greater obesity. It is important to remember that *refined* carbohydrates are the problem. Foods high in fiber and complex carbohydrates (e.g., grains, vegetables, and fruits in the form Nature made them) do not cause a spike in blood glucose because the fiber slows the rate of absorption.

Under the scenario described above, the solution to obesity seems self-evident; strictly limit the consumption of carbohydrates. This will cause the levels of blood glucose to drop, which in turn will signal the pancreas to stop releasing insulin into the bloodstream. With less insulin being released, the body is free to burn its own stored fat and pounds start melting away. As long as carbohydrates are restricted, you can eat unlimited amounts of fatty foods with no fear of gaining weight. Even if fat is present, insulin production is very low, so there is no signal to start stashing it away.

The best-known of the hormonal or low-carbohydrate advocates is the late Dr. Robert C. Atkins, who wrote *Dr. Atkins' Diet Revolution* (1972), perhaps the most successful weight loss book of all time. Various editions of this book sold more than 15 million copies. Atkins attracted a multitude of followers and eventually achieved cult-like status. So great was his influence that a diet high in fat and low in carbohydrates is now commonly referred to as the "Atkins diet." His message: high-carbohydrate foods such as bread, rice, and potatoes lie at the heart of modern illness; the very act of eating them alters the chemistry of the body, dooming us to overeat and become obese; and in many cases, diabetes and heart disease follow closely behind.

Atkins had reason to believe in the low-carbohydrate regimen; over many years, he had used it to successfully treat thousands of obese and diabetic patients in his New York City medical practice. Atkins, like Blake Donaldson before him, noted that patients on a low-carbohydrate diet, in addition to losing weight, often saw symptoms of heart disease, hypertension, diabetes, and other metabolic illnesses abate as well. He repeatedly challenged his critics to come to his office in New York City and review his patient files.

Dr. Eric Westman heads up the Lifestyle Medicine Clinic at the Duke University Medical Center in Durham, North Carolina. In 1998, Westman became intrigued when one of his patients declared that he had lost 20 pounds in two months while eating mostly steaks and salads. This prompted Westman to read up on the low-carbohydrate diet and to meet with Dr. Robert Atkins at his office in New York City. He persuaded Atkins to fund a pilot study to determine if the diet was effective and safe. The study, which involved 50 patients and lasted for six months, demonstrated that patients did lose weight on the diet and that it appeared to be safe, resulting in improvements in cholesterol profiles and other health markers (Westman et al. 2002).

Westman then visited at least four other physicians around the country who, like Atkins, were using high-fat, carbohydrate-restricted diets in their medical practices. All of these physicians reported that restricting carbohydrates was an effective therapy for treating obesity and related metabolic disorders. Westman returned to Duke and began treating patients with the diet, and he continues to do so. In addition, he and his colleagues have carried out a number of clinical trials around the country, all of which confirmed the efficacy of carbohydrate restriction in treating obesity and diabetes, and presumably, the future incidence of heart disease; for examples, see Vernon et al. (2003), Volek et al. (2009), and Westman et al. (2006).

Two Patients

Imagine two extremely sick middle-aged men. In addition to being morbidly obese, both are diabetic, hypertensive, and have very alarming cholesterol and triglyceride numbers. Both are taking large doses of medication to control their blood pressure, statins to lower their cholesterol, and daily insulin injections to normalize their blood sugar. Each has a body mass index (BMI) greater than 40, and each is continuing to gain weight. The various prescriptions they take are not working well, so the dosages must be increased periodically and/or new medications must be added. Recognizing that conventional drug therapy is no longer working for these patients, their physicians, as a last resort, refer them to lifestyle clinics. Such clinics emphasize dietary and behavioral solutions to obesity and cardiometabolic disease.

One of the men lives on the west coast, and his doctor refers him to the Lifestyle Medicine program run by Dr. Dean Ornish, professor of clinical medicine at the University of California, San Francisco. Ornish is perhaps the best known advocate for high-carbohydrate, fat-restricted diets. He has written four widely-read books on the use of very low-fat diets to treat obesity and related metabolic disorders. Although the message and tone vary somewhat from book to book, the underlying rationale, as stated in his first book, is the same, "when you go from a high-fat to a low-fat diet, you eat fewer calories without having to eat less food" (Ornish 1993).

Ornish has the singular distinction of being a diet guru who was inspired by a religious guru. In 1972, ill and suffering from depression, he dropped out of Rice University in Houston. While at his parents' home in Dallas, he was introduced to a Hindu swami or guru, who at the time was instructing Ornish's older sister in meditation and relaxation exercises. The young college student took advantage of the opportunity and sought spiritual and health counseling from his sister's mentor. The swami advised him to become a vegetarian, exercise, meditate, practice yoga, and always seek to help others. The young man took this advice to heart, completed his medical studies, and the rest, as they say, is history (Kolata 1998).

Upon reporting to Ornish's Lifestyle Medicine clinic, our fictional patient will be put on a very low-fat, high-carbohydrate diet, one in which fat provides less than 15 percent of calories, while complex, unrefined carbohydrates provide more than 60 percent. The patient will eat a lot of whole grains, legumes, and fresh vegetables. Such a diet has been used successfully to treat obesity, diabetes, and hypertension in thousands of people. If our imaginary patient follows the program, he most likely will see enormous improvement within only a few months, and within a year will be well on his way to metabolic health. This should not surprise us, since the Ornish diet is simply a return to the past — to the grain and legume based diets eaten by our agricultural ancestors for thousands of years.

But in order to reprise this ancestral diet, Ornish also must restrict the amounts of sugar and other refined carbohydrates, which became a prominent part of western diets during the 20th century. In his 1993 book, *Eat More, Weigh Less*, Ornish explains that, while you can eat as many total carbohydrates as you want, it is important to restrict certain kinds of carbohydrates. By strictly limiting sugar and refined carbohydrates, which are hallmarks of the industrial food system, Ornish is returning his patients to the minimally-processed, plant-based diet eaten by much of the world's agricultural population prior to the 20th century. He writes,

> When you eat a lot of sugar and refined carbohydrates, you consume a lot of calories without feeling full. Also, you provoke an insulin response, which helps convert those calories into fat. However, a diet high in fiber and complex carbohydrates doesn't cause a spike in blood glucose because the fiber slows the rate of absorption.

Let us now turn to our other fictional patient, who lives on the east coast of the United States and is referred by his primary physician to Dr. Eric Westman, director of the Lifestyle Medicine Clinic at Duke. Since 2006, Dr. Westman has successfully treated thousands of severely obese patients suffering from a variety of cardiometabolic diseases. But here is the really interesting part. The dietary program prescribed by Dr. Westman at Duke is the exact opposite of the one prescribed by Dean Ornish. Instead of a high-carbohydrate, nearly vegetarian diet, Westman uses a high-fat, nearly carnivorous diet, or what commonly is referred to as the *keto diet*.

Westman's patients are free to eat as much meat, fish, cheese, and other animal foods as they like. They also are allowed to eat such foods as lettuce, spinach, broccoli, cabbage, and tomatoes, but sugary fruits and starchy vegetables are largely off the menu. Needless to say, refined sugar is strictly forbidden. Using this diet, Dr. Westman and others have successfully treated obesity, diabetes, and hypertension in thousands of people. If our imaginary patient follows this program, he soon will begin to lose weight, and within a year or so his metabolic health most likely will have normalized. This should not surprise us, because, as in the case of the Ornish program, the keto diet is simply a return to our nutritional past — but to a different past. Whereas Dr. Ornish returns his patients to the grain and legume-based diets of their agricultural ancestors, Westman resurrects the more ancient hunter-gatherer diet.

In 2000, Australian and American scientists evaluated the diets of 229 hunter-gatherer groups that survived into the 20th century long enough to have their food consumption assessed by anthropologists. According to their analysis, "whenever and wherever it was ecologically possible," hunter-gatherers consumed "high amounts" of animal foods (Cordain et al. 2000). They found that 20 percent of the hunter-gatherer populations studied lived almost entirely by hunting or fishing, getting more than 85 percent of their food energy from meat or fish; some groups got nearly 100 percent. Only about 15 percent of the groups got more than half of their calories from plant foods, and none was exclusively vegetarian. It is clear that humans are able to live healthy lives and successfully rear children on a diet made up largely of animal foods. On average, the 229 hunter-gatherer groups got about 65 percent of their calories from animal foods and about 35 percent from plant sources.

The standard keto diet, as prescribed by Eric Westman, consists of more than 60 percent fat and only about 10 percent carbohydrates. In contrast, the standard low-fat diet, as prescribed by Dean Ornish, derives more than 60 percent of food energy from carbohydrates and only 10 to 15 percent from fat. The fact that both of these contrasting regimens work reflects the dietary history of our species. From roughly 200,000 years ago until about 10,000 years ago, many of our ancestors consumed diets that were, in essence, the one currently prescribed by Dr. Westman at Duke and by other physicians around the world. Similarly, the plant-based, high-carbohydrate diet currently prescribed by Dr. Ornish and other like-minded physicians is, in essence, the same diet eaten by most of the human race following the adoption of agriculture beginning about 10,000 years ago.

It is revealing to examine the books on display in the health and diet section of any large bookstore. You will see a variety of books proclaiming that the only way to prevent obesity, diabetes, and heart disease is to eat a diet that severely restricts carbohydrates. But on the same shelf there usually are an equal number of books proclaiming just the opposite — that the only way to remain free of these modern ailments is to eat a diet that severely restricts dietary fat. To some, this might seem like walking into the geology section of a bookstore and finding that about half of the manuscripts proclaim that Earth is flat, while the other half proclaim that it is round. But in the case of diet, the contradiction is only apparent. This is the hallmark of a classic paradox, which is a statement or idea that seems contradictory or opposed to common sense, but on close inspection, turns out to be true. Comprehending the nature of this paradox provides us with the insight needed to fully understand the modern epidemic of obesity and metabolic disease.

Obesity, especially abdominal obesity, is the first and most obvious manifestation of metabolic disease, and the harbinger of worse things to come. When you notice the beginning of abdominal obesity in young people, as is so often the case today, you most likely are seeing children with diabetes, hypertension, and even heart disease in their futures. In order for someone to become obese, two things must happen. First, he or she must take in excess calories on a more or less daily basis over an extended period of time. This means that obesity has an energy density or *thermodynamic* component. Second, something must prompt the body to store these excess calories as fat rather than burn them off. The hormone that does that is insulin. This means that obesity also has a *hormonal* component.

A diet that strictly limits fat consumption, such as the one prescribed by Dean Ornish, attacks obesity by reducing the energy density or caloric content of the foods eaten. As discussed earlier, a gram of fat contains twice as many calories as a gram of carbohydrate or protein. The large amounts of carbohydrate eaten on a low-fat diet will stimulate insulin production, so the hormonal component will be there. But such a diet provides few excess calories to be stored; and even if too many calories are taken in by overconsuming starchy foods such as rice, potatoes, or beans, the body is very inefficient at storing excess carbohydrates as fat.

In contrast, the high-fat or keto diet, such as the one prescribed by Eric Westman at Duke, attacks obesity from the hormonal side of the equation. A diet high in fat is by definition a diet low in carbohydrate, and by severely limiting carbohydrate consumption, such diets drastically curtail the production of insulin. Under these conditions, even if excess calories are available in the form of dietary fat, very little of it will be stored in the body. A major reason for this is the satiating effect of high-fat foods; it is hard to overeat on such a diet. Since there is little insulin to drive fat into storage, nearly all of it is available for the body to use for fuel as it is consumed. This puts a brake on overeating, fat storage, and weight gain. So both the high-fat and the high-carbohydrate approaches work, each in its own way.

It is this duality that has made modern human obesity such a vexing problem, and at the same time such a fascinating one. Because they have assumed that there must be one and only one cause of metabolic disease, scientists have been engaged in a pointless debate for more than a century. One side has argued that excess dietary fat is the villain, while the other side has insisted with equal fervor that carbohydrates are the problem. Both sides have assumed that it had to be one or the other. This mindset probably reflects the influence of pioneering scientists such as Louis Pasteur and Robert Koch, who determined more than a century ago that for each bacterial or viral disease, there is one organism responsible — one disease, one cause.

During the 20th century, those who began to study metabolic diseases such as obesity and diabetes tacitly assumed, incorrectly as it turns out, that the infectious disease model would apply to metabolic diseases as well. This fixation on identifying a single cause for obesity and associated ailments led to the fat-versus-carbohydrate debate, which has lasted more than a century, and sadly, is still going on. Scientists undertaking nutrition research during the 20th century were basing their reasoning on a faulty assumption. As a result, in the wise words of a former colleague, they allowed themselves to become "trapped inside a bad question."

Chapter 9. Sweet and Salty

As discussed in an earlier chapter, industrial foods such as cakes, cookies, ice cream, pastries, and candy represent a form of food which did not exist on Earth until very recent times. To my knowledge, no natural food contains both large amounts of sugar and large amounts of fat. Fruits are sweet but contain almost no fat, while nuts and certain animal parts contain a lot of fat, but almost no sugar. Nature was wise to keep these two apart. Both fat and sugar, if eaten alone, are largely self-limiting. But mixing them together can be a real problem.

Sugar makes fat taste better, while at the same time, added fat keeps foods containing as much as 70 percent sugar from tasting too sweet. Evolution has not prepared us for this combination. One scientist found that a mixture of 40 percent fat, 40 percent sugar, and 20 percent refined flour made rats enormously fat in a very short time. Large amounts of fat alone were not sufficient; as one scientist observed, "It took the addition of sugar to light the fuse" (Arnot 1997). Many Americans are becoming obese and unhealthy by making these beguiling fat-sugar combinations a large part of their diets.

Table 9.1 shows how sugar and fat consumption changed in the American diet between 1950 and 2000. Data are from the Agricultural Fact Book, 2001-2002 (USDA 2002). During the 20th century both fat and sugar consumption rose to well above evolutionary norms, but we should be especially concerned about the increased consumption of sugar.

TABLE 9.1. SUGAR AND FAT IN THE AMERICAN FOOD SUPPLY (POUNDS/CAPITA/YEAR)

Nutrient	Year 1950	Year 2000	Percent change
Total fats and oils	114	128	+12%
Total sugars	110	152	+38%
Corn sugars	11	86	+682%

The relative increase in sugar consumption from 1950 to 2000 was more than three times that of fat. By 2000, the typical American was getting nearly 700 calories daily just from pure sugar and corn syrup, drinking much of it in the form of carbonated soft drinks. In 1947, the first year for which soft drink consumption data are available, each American consumed about ten gallons. But by 2000, consumption levels had increased more than five times, with the typical American drinking more than 50 gallons of these sugary concoctions (USDA 2002).

In 1950, the standard soft drink size was 6.5 ounces, but by 1960 the 12-ounce can had become the standard. By 1990, increasing numbers of people were buying Coke and other soft drinks in 20-ounce bottles. Now, you can buy a 64-ounce soft drink containing up to 700 kilocalories at almost any fast food restaurant or convenience store. Advertisements in the 1950s described a 12-ounce serving of soft drink as "king size." Now fast food restaurants commonly refer to a 32-ounce drink as "medium size." Nearly one fourth of Americans get 200 kilocalories or more daily from soft drinks alone, which now provide nearly ten percent of the nation's total calorie intake. It is estimated that more than 60 percent of American children consume sugary drinks on a daily basis; for poor children it is more than 75 percent (Koma et al. 2019). A recent review article by DiNicolantonio et al. (2016) emphasized the health consequences of such high sugar consumption.

> A diet high in added sugars has been found to cause a 3-fold increased risk of death due to cardiovascular disease... To reduce the burden of CHD [coronary heart disease], guidelines should focus particularly on reducing intake of concentrated sugars, specifically the fructose-containing sugars like sucrose and high fructose corn syrup in the form of ultra-processed foods and beverages.

A Brief History of Sugar

The first known shipment of sugar reached Europe in AD 996, but it remained rare on the continent for the next 500 years. In 1288, the kitchen staff of King Edward I is reported to have used more than 6,200 pounds of sugar (Mintz 1985). But sugar was too expensive to become a common food in Europe until the 1600s, when cane plantations were established in the Caribbean. Shakespeare's play, *The Winter's Tale*, written in the late 1600s, includes a shopping list to feed 24 people at a country feast. Among the items listed are three pounds of sugar.

The rich merchants and the nobility of England began to consume large amounts of sugar during the 1600s, and within only a few decades they began to show the effects. As described in Chapter 1, the National Portrait Gallery in London's Trafalgar Square contains portraits of all of the English monarchs from the 14th century on. King Henry VII, painted near the end of the 15th century, is shown with a slight double chin, but true obesity does not make an appearance until the last half of the 1600s, when Charles II is portrayed as being notably obese. By the early part of the 1700s, every member of the nobility pictured, lord or lady, had become overweight or obese (Galton 1976). The rapidity with which obesity followed the accelerating consumption of sugar during the 1600s is striking.

By 1700, the average British citizen was consuming nearly five pounds yearly of what still was a rare delicacy (Farb and Armelagos 1980). Sugar still was so expensive that many households kept it under lock and key. By 1850, the price of sugar had declined significantly, and the average British citizen was consuming more than 20 pounds each year (Farb and Armelagos 1980). Now, a little more than 150 years later, the average British and American citizen is consuming more than 100 pounds of sugar annually, most of it in the form of artificially sweetened foods and drinks.

USDA market data show that in 1999, 150 pounds of sugar, much of it in the form of high fructose corn syrup, were sold for every man, woman, and child in the United States (USDA 2002). This should not surprise us. Absent a deliberate exercise of will, our instincts lead us to consume foods that provide energy in its tastiest, most easily available form. And sugar is the ultimate in this regard, consisting of pure calories. By the end of the 20th century, the average American was getting about 500 kilocalories daily just from sugar and corn syrup.

In 1950, each of us took in a little over 75 pounds of caloric sweeteners, which are concentrated sugars and syrups derived mostly from sugar cane, sugar beets, and corn. By 2000 this had increased to almost 110 pounds per capita. Interestingly, much of this increase took place after 1980, following the introduction of high fructose corn syrup. We were consuming no high fructose corn syrup prior to 1970, but by the mid-1980s each of us was consuming nearly 20 pounds of this new product, mostly in soft drinks and sweetened convenience foods. Per capita consumption of high fructose corn syrup had risen to more than 45 pounds by 2000, and Americans were taking in 350 more kilocalories each day than they had in 1983, thanks largely to this new product (Gerrior and Bente 1997).

By the end of the 20th century, many Americans were consuming their body weight or more in sugar every year and, as a result, our nation had grown increasingly obese and diabetic. That should not be a surprise; prompting animals to manufacture fat and store it away for later use is precisely what Nature designed sugar to do. An early account of the fattening effects of sugar comes from an English doctor in the mid-1800s. He describes how consuming a lot of sugar during the harvest affected both animals and men on sugarcane plantations (Tanner 1869).

> In sugar-growing countries both the [men] and cattle employed on the plantations grow remarkably stout while the cane is being gathered and the sugar extracted. During this harvest the saccharine juices are freely consumed; but when the season is over, the superabundant adipose tissue is gradually lost.

Sugar and Seasonality

More than four billion years ago, a Mars-size body called *Theia* struck the young Earth a glancing blow, knocking it permanently off of its axis and leaving it tilted at an angle of 23.5°. As a result, areas of our planet away from the equator have distinct winter and summer seasons. Plant species outside the tropics have adapted to this seasonality by growing and photosynthesizing during the summer and then shedding their foliage and going dormant during the winter.

In addition to winter and summer, there is another kind of seasonality. Rain does not fall evenly throughout the year in many parts of the world. Instead, because of warm and cold ocean currents, different heating and cooling rates of land and water, the effects of high mountain ranges, and perhaps other factors as yet unknown, many land areas have distinct wet and dry seasons. Plant species have adapted to this kind of seasonality also, often by shedding their leaves and becoming dormant during the dry seasons.

Animals that live on land, including humans, also evolved strategies to cope with seasonality, to live through winters or dry seasons when food was less plentiful. One strategy was to store extra energy as body fat during the warm or rainy season when calorie-rich food was more abundant, and this is where sugar played a role. The ability to turn part of the sugar we eat into fat and store it in our bodies for later use evolved very early in hominids. This adaptation has been carried down through many generations because of its survival value, especially as many of our species left the tropics and migrated into colder regions.

Imagine a small group of hunter-gatherers living somewhere in the Middle East 40,000 years ago. It is early summer and they are somewhat lean from the cold winter that ended just weeks earlier. They have lived in the same territory for generations and know it well; they know every source of water, every place where animals gather, every cluster of nut-producing trees, every fruit tree, and every berry patch. Not only do they know the location of every fruit tree and berry patch, they know when the fruits flower and when they ripen. Nearly all foods are more plentiful during the summer than during the winter — but the main difference is the availability of fruits. During the warm season, the foraging band visits every known source of these delicious, high-energy foods, and returns to some of them again and again as new fruit ripens.

They do not know it, but on nearly every visit to one of their favorite berry patches or fruit trees, they convert small amounts of the fruit sugars they consume into fat and store it in their bodies. Using the USDA nutritional database FoodData Central at fdc.nal.usda.gov, I determined the typical amounts of sugar in ten commonly eaten fruits — apples, blackberries, blueberries, cherries, grapes, pears, persimmons, plums, raspberries, and strawberries. Since we are eating these fruits today, it is very likely that our ancestors ate their ancestors many generations ago. On average, these fruits consist of about 85 percent water, 10 percent sugar, and 3 percent fiber. Small amounts of protein, minerals, and even a little fat make up the remaining 2 percent. About half of the sugar in these fruits is glucose, while the other half is fructose. A small amount of the sugar is actually sucrose, but every sucrose molecule consists of a fructose molecule linked to a glucose molecule. Our bodies quickly cleave these two sugars apart, so from a functional standpoint, we can view fruit sugar as consisting of about half fructose and half glucose.

Let us return to our band of hunter-gatherers and consider how their bodies metabolize each of these sugars taken in during the time they spend at one of their favorite berry patches. Assume that, a few hours before deciding to stop for berries, our small group of foragers had visited a grove of nut trees where each of them had eaten some nuts. By the time they reach the berry patch, their bodies already have begun digesting the nuts, releasing fat into their bloodstreams.

As our hunter-gatherers begin eating fruit, sucrose, glucose, and fructose all start flowing into their bloodstreams. The presence of glucose in the blood signals the pancreas to start releasing insulin. The insulin spreads throughout the body and signals the muscle cells to stop burning fat and to start taking in glucose and burning it instead. Insulin also directs the fat cells to hold on to any fat they already have in storage and to start taking in additional fat and storing it also.

Unlike glucose, the fructose in fruit goes directly to the liver, where it is easily converted into fat. Some of this fat is stored in the liver, while some is released into the blood in the form of triglycerides. The insulin in the bloodstream signals the fat cells to break down these triglycerides into fatty acids, which are small enough to pass through the cell membrane and into the cell. Once inside the cell, the fatty acids are reassembled; and the fat synthesized from fructose is stored along with that coming from the nuts eaten earlier.

Because fruit contains so much fiber — one gram of fiber for every 3 grams of sugar in our sample, eating it does not cause a sudden surge of sugar into the bloodstream. Instead, the sugar is absorbed slowly, causing only a gradual rise of glucose in the blood, a gradual release of insulin from the pancreas, a gradual transformation of fructose to fat in the liver, and a slow but steady movement of fat into the fat cells. One can imagine each of our hunter-gatherers storing away a few extra grams of fat every day or so. By the end of summer, each of them will have built up a fat surplus of 5 to 10 pounds, enough stored energy to get them through any lean days they might encounter during the coming winter.

This well-regulated process served humanity well throughout most of our time on Earth. But in 20th century America, this ability to turn fructose into fat and to store fat in our bodies started to become a curse. The Economic Research Service (USDA 2002) estimated that in 2000, the average American consumed about 32 teaspoons (135 grams) of added sweeteners every day. These include the sugar added to cereal or tea, and the sweeteners that the food industry adds to carbonated drinks such as Coke, and to cakes, cookies, canned soups, pickles, and a wide variety of other commercial foods.

In 2000, about 280 pounds of fruit or fruit juice were sold for every man, woman, and child in America (USDA 2002). We know that fruit has a high spoilage rate, so let us assume that only half of it (140 pounds per year) was eaten. This means that, in addition to added sweeteners, the average American in 2000 was consuming about 42 teaspoons of sugar each day from fruit or fruit juices, bringing the total daily sugar consumption to 72 teaspoons per capita.

Regrettably, we are not yet finished; there is yet another major source of sugar in the modern American diet. According to USDA, about 150 pounds of wheat flour enter the food system each year for each man, woman, and child in the United States. About 50 pounds of corn and rice products also enter the system, bringing the total marketing of grains in our economy to 200 pounds per capita. This is equivalent to about 145 pounds of starch, and starch consists of long chains of glucose molecules. The starch from foods such as bread and white rice begins breaking down into sugar as soon as we start chewing it and before long glucose molecules are entering our bloodstream. White rice and bread are said to have a high *glycemic index* because eating them raises the glucose content of the blood so rapidly.

If we assume that about 60 percent of the flour, corn, and rice entering the food system actually gets eaten, then grain consumption adds 90 pounds of glucose per year to the diet of the average American — an additional 26 teaspoons of sugar every day. Altogether, fruits, fruit juices, added sweeteners, and grains provide the average American with nearly 100 teaspoons of glucose daily. In order to match that, the average hunter-gatherer would have needed to eat about two-and-a-half pounds of fruit every day, year round.

A large number of health-conscience Americans eat very little commercially added sweeteners, and many also limit the amount of bread and fruit they eat. Since such individuals are consuming much less than the average amount of sugar, it follows that others among us must be consuming much more. There are significant numbers of Americans, notably those drinking a lot of sweetened beverages, who are taking in the equivalent of more than 150 teaspoons of sugar every day, or about 500 pounds per year. Considering this, it should be no surprise that the percentage of obese Americans has nearly tripled since 1970, increasing from about 15 percent to more than 42 percent. Nor should it surprise us that one-third to one-half of all American children born in this century can expect to become both obese and diabetic at some point in their lives.

Someone who consumes large amounts of sugar on a daily basis almost inevitably develops a condition called non-alcoholic fatty liver disease, of NAFLD for short, as opposed to alcoholic fatty liver disease, which develops in people who consume a lot of alcohol on a regular basis. The liver metabolizes ethyl alcohol and fructose in much the same way, and overconsuming either can cause liver disease. There is a reason that the human liver can break down both fructose and alcohol and turn both of them into fat. An earlier section of this chapter emphasized that being able to make fat from fructose and store it in the body for later use had survival value for our hunter-gatherer ancestors.

The same is true of alcohol. Being able to metabolize ethanol without becoming ill enabled early foragers to take advantage of fruit that had fallen to the ground and fermented, and it is much easier to pick fruit up off of the ground than to climb a tree. But as in the case of sugar, Nature could not anticipate that clever humans one day would learn to make very concentrated solutions of alcohol and that many of them would start to consume large quantities of it on a daily basis. Again, as with the ability to metabolize fructose, an adaptation that served our species well in the distant past has now become a curse to many.

The K Factor

Early in the 20th century, scientists at the United States Geological Survey analyzed more than 5,000 rock samples from around the world and calculated the average elemental composition of the Earth's crust. They found that the rocks beneath our feet consist overwhelmingly (more than 98 percent by weight) of just eight chemical elements (Clarke 1924). We will focus on just two of them — sodium (Na) and potassium (K). By weight, Earth's surface rocks consist of about 2.7 percent sodium and 2.5 percent potassium. As the upper part of the crust is broken down to form soil, both sodium and potassium are released from the rock structure; they then take very divergent paths.

Although terrestrial rocks contain about equal amounts of the two elements, seawater contains about 28 times more sodium than potassium, and the reason for this is quite simple. Most of the sodium released during rock weathering is quickly leached from the land and carried away to the sea by streams and rivers. In contrast, most of the potassium stays on the land. A lot of it is stored in the soil, much of it as part of a type of mineral called mica. Mica is very distinctive in appearance because large piece of it can be peeled apart into thin, flexible sheets — and potassium, in the form of positively charged K^+ ions, is the "glue" that holds these sheets together.

Mica serves as a storage depot for potassium in the soil. During the growing season, some of the mica breaks down and the potassium atoms holding the sheets together are released to be taken up by plant roots. This is fortuitous, because plants take up a lot of potassium and store a lot in their tissues. Trees are a good example. The name potassium is derived from "potash," referring to wood ashes. People learned centuries ago that they could concentrate potassium salts by burning wood and then mixing the ashes with lye and boiling the mixture in a pot, hence "potash." The result was a valuable industrial commodity, useful for making such things as glass, soap, and gunpowder, and for dyeing and bleaching fabrics.

But we are more concerned with the amount of potassium in the plant parts and plant tissues that we eat. More specifically, we are interested in the amount of potassium relative to the amount of sodium, commonly referred to as the potassium–sodium ratio or K factor. Making sure that the balance between potassium to sodium, or K factor, remains within the evolutionary norm is critical to human health. But what is that evolutionary norm? Table 9.2 shows the ratio of potassium to sodium (K/Na) of some traditional hunter-gatherer and peasant foods. The data are from the USDA nutrient database (fdc.nal.usda.gov).

Table 9.2. Potassium to sodium ratios (K to Na) of some traditional foods

Food	K to Na ratio
Shellfish	1.5 to 1
Ocean fish	3 to 1
Freshwater fish	4 to 1
Meat and fowl	4 to 1
Mushrooms	30 to 1
Wheat, corn, rice	40 to 1
Beans, peas, lentils	70 to 1
Fruits	130 to 1
Nuts	180 to 1

It is clear from Table 9.2 that, for most of human history, our species ate foods that contained a lot more potassium than sodium. In other words, traditional hunter-gatherer and peasant diets had a high potassium to sodium ratio, or K factor. This was the evolutionary norm for our species. For comparison, Table 9.3 shows the potassium to sodium ratios of some representative restaurant and convenience foods that now make up much of the American diet.

TABLE 9.3. POTASSIUM TO SODIUM RATIOS (K TO NA) OF SOME MODERN FOODS

Food	K to Na ratio
Fast food hamburger	1 to 3
White bread	1 to 4
Fast food beef burrito	1 to 4
Hot dog	1 to 4
Fast food pizza	1 to 4
Cookie	1 to 4
Bagel	1 to 5
Fast food fried shrimp	1 to 5
Corn flakes	1 to 5
Frozen chow mein dinner	1 to 8

A comparison of Tables 9.2 and 9.3 is revealing. The diet eaten by humans throughout most of history contained four to five times more potassium than sodium. Then, mostly during the last century, this evolutionary norm was totally reversed in most industrialized nations. Even foods that do not taste salty, such as bread, rolls, cookies, cakes, and breakfast cereals, contain at least four times more sodium than potassium, and this is by design.

A few years ago, I saw a promotional video put out by a major food company. In the video, a company spokesman is poetic in his praise of salt, proclaiming, "You might be surprised by what foods are enhanced by its briny kiss." This spokesman had a point; it is hard to get through the day without experiencing the "briny kiss" of salt many times over. The cereal and toast we eat for breakfast, the soup and crackers we enjoy for lunch, and the fast food hamburger we get at the drive through after work — all have been laced with generous additions of salt.

Without a lot of added salt, many processed foods such as cereals, crackers and processed meats would taste like "wet cardboard" or "damp dog hair" (Moss 2010). Here is the rather unpalatable truth: if it were not for generous infusions of salt, many modern foods would simply taste bad. So salt is indeed "life enhancing," but mostly for the food industry. Without it, the huge sales and impressive profits they enjoy would not be possible. Salt is an all but indispensable part of the modern factory food system.

But salt might not be so life enhancing for the rest of us. According to Michael Jacobson, executive director of the Center for Science in the

Public Interest, "Salt is the single most harmful element in our food supply, silently killing about 100,000 people each year" (Hellmich 2010). Really scary comments, but could salt really be that bad? The American Medical Association (AMA) seems to think so. A few years ago, the AMA requested that the Food and Drug Administration revoke salt's long-time designation as a substance that is "generally recognized as safe," arguing that we are eating way too much of it and killing ourselves in the process. The AMA is not alone; most health professionals are convinced that consuming too much salt causes hypertension, which in turn leads to the death of thousands of people each year from stroke or heart attack.

Here is why the proper balance between dietary sodium and potassium is so important. The cells of your body, in addition to their other duties, function as little power plants that generate electric current, but the electricity generated by our cells is not carried by electrons; instead, it is carried by positively-charged sodium atoms (Na^+ ions). Each cell can produce almost one-tenth of a volt across its thin outer membrane. This does not seem like a lot, but relative to its size, a human cell can produce as much electrical energy as a nuclear power plant. An electric eel, by lining up millions of its sodium-potassium pumps in just the right way can generate enough voltage to give you a nasty shock.

Sodium-potassium pumps move charged sodium atoms (Na^+ ions) out of the cell, while moving charged potassium atoms (K^+ ions) into the cell. Overall, about three Na^+ ions are moved out for every two K^+ ions that are moved in. It is the net flow of Na^+ ions out of billions of cells that generates the body's electrical current. This process also assures that the cell has a high K factor — that is, that there is always more potassium than sodium inside our cells. An indication of how important this is — when cells are at rest, about 25 percent of the energy they use is to keep their sodium–potassium pumps functioning.

Everything works fine as long as most of the foods you eat contain more potassium than sodium, which was the case throughout most of human history. As shown in Table 9.1, anyone who consumes traditional hunter-gatherer and/or peasant foods will be eating a diet with a high K/Na ratio or K factor, which keeps the sodium–potassium pumps charged up and generating power. At the same time, it ensures a proper balance between sodium and potassium, both inside and outside the cell. Our species evolved to consume a diet that contains four to five times more potassium than sodium.

But during the last century, people in America and other industrialized nations began replacing real foods with highly-processed factory foods, many of which were made palatable only by adding lots of salt. In just a few generations, millions of humans went from eating a diet containing four to five times more *potassium* than *sodium* to one containing four to five times more *sodium* than *potassium*. An evolutionary norm was totally reversed, and that has not been good for human health.

A diet containing too much sodium and not enough potassium prevents the sodium–potassium pumps from working effectively, and excessive amounts of sodium start to build up in cells. Along with sodium, excessive amounts of calcium also begin to accumulate, and the poorly functioning sodium–potassium pumps are unable it move it out — and this is a problem. Even a small increase in the amount of calcium in muscle cells causes them to contract. This is a special problem with the muscle cells surrounding small arteries. As the muscle cells contract in response to the increased calcium, they begin to squeeze the small arteries, causing them to narrow and restrict the flow of blood. The heart then must work harder to push blood into the constricted arteries, causing a rise in blood pressure (Moore 2001).

According to this rationale, people who live in modern societies but manage to avoid salt-infused industrial foods by eating a vegetarian or plant-centered diet should have low rates of hypertension — and a number of studies have shown this to be true. For example, high blood pressure was found to be very rare among vegetarians in Boston (Sacks et al. 1974) and among Trappist monks in Holland and Belgium (Groen et al. 1962). The same was true of Seventh-Day Adventists in Australia (Armstrong et al. 1977) and vegetarians in Israel (Ophir et al. 1983). Only 1% to 2% of these groups suffered from high blood pressure, compared with 30% to 40% of the surrounding populations. A vegetarian or plant-centered diet is almost by definition low in sodium and high in potassium. Unfortunately, although highly suggestive, such observations do not really prove that too much sodium relative to potassium is a primary cause of hypertension. It could be something else related to food, such as lack of fiber or a certain vitamin deficiency.

However, there is good evidence from Finland showing that a high ratio of sodium to potassium in the diet is indeed a major cause of hypertension. During the 20-year period from 1972 to 1992, the entire nation of Finland took part in a revolutionary nutritional experiment. At the urging of the Finnish government, the food and restaurant industries agreed that, whenever possible, they would replace regular salt (sodium chloride) in commercial foods with a mixture called *Pansalt*, in which 43 percent of the sodium chloride (NaCl) was replaced with non-sodium compounds, mostly potassium chloride (KCl). The Finns were not banning industrial foods or adding less salt to them; they simply were changing the chemical composition of the added salt to create a healthier mix, i.e., one that included more potassium and less sodium. The results were impressive. In only a few years, there was a significant decrease in blood pressure throughout the country. More to the point, deaths from stroke declined by 60 percent, while mortality from heart disease went down 55 percent in men and 68 percent in women (Karppanen and Mervaala 1996).

By the end of the 20th century, fat consumption in the United States had risen well above the evolutionary norm, and sugar consumption had become many times greater than the level for which evolution had prepared us. At the same time, we began to consume large amounts of salt, or sodium chloride, disrupting the dietary balance between sodium and potassium. We evolved to eat much more potassium than sodium, but within just a generation or so, this evolutionary norm had totally reversed. In addition to eating more fat, more sugar, and more salt than evolution had prepared us for, we also became increasingly sedentary. According to one author, "It appears that by 1977 Americans had reached activity levels that were approaching an all-time minimum for the human species" (Bennett and Gurin 1982). The next chapter describes how America became a sedentary nation and why this is such a problem.

Chapter 10. Sedentary Nation

During the 20th century, Americans greatly increased the energy density of their diets, mainly by eating a lot of fat, a lot of refined carbohydrates, and a lot of sugar. At the same time, they began to consume ever larger amounts of added salt, leading to an unhealthy ratio between dietary sodium and potassium. While eating a lot of fat, sweet, and salty foods that were almost guaranteed to make us obese and sick, we compounded the problem by becoming increasingly sedentary. By the second half of the 20th century, automation had created a society in which few of us had to do any real physical work. A long-term study of San Francisco longshoremen illustrates this. In the early 1950s, about 40 percent of the men were doing "heavy work," expending roughly five kilocalories each minute (about 2,300 per day) in lifting and carrying. Just ten years later, as a result of mechanization, only 15 percent still were performing heavy labor; and by 1972, with the introduction of containerized shipping, this had declined to five percent (Paffenberger and Hale 1975).

Tractors and Farmers

Henry Ford began selling a small, rubber-tired farm tractor during World War I, when there was a shortage of horses. It sold well and other companies soon entered the market. Then in 1922, the power take-off was invented. This is a deceptively simple device, just a rigid steel shaft with raised surfaces enabling it to fit like a key into another piece of machinery. The power take-off transfers power from a tractor's engine to a planter, mower, corn picker, or other implement being pulled behind. Engineers responded by designing all kinds of farm equipment to work off of a power take-off.

The results were revolutionary. For 10,000 years, humans and animals had supplied the labor to plant, till, and harvest crops; but by the middle of the 20th century, tractors and tractor-driven equipment had largely replaced humans, horses, and mules (White 2000). The tractor and power take-off altered the landscape of America forever. In time, millions of horses and mules, as well as millions of farm workers, were no longer needed, and waves of people began leaving the land. The great American heartland emptied out, a process that took several decades. In 1920, farmers and farm laborers accounted for more than 25 percent of the nation's workforce. This declined to only 12 percent in 1950 and then to less than five percent by 1970. Today slightly less than one percent of Americans are full-time farmers or farm workers.

As agricultural areas lost people, the cities began growing rapidly, both from the influx of displaced farm workers and the arrival of immigrants from overseas. In 1900, only four American cities — New York, Chicago, Philadelphia, and Boston, had populations of over one million. Phoenix was then a small town with a population of only 5,000 people or so; Miami, with fewer than 2,000 people, had just been incorporated. Las Vegas was so small in 1900 that the Census Bureau did not even recognize it as a town. But these and many other cities grew steadily during the 20th century. Between 1900 and 2000, our nation added about 200 million people, and most of them became city dwellers. We now have more than 35 urban areas (central cities plus densely settled suburbs) with populations greater than one million. The 20th century saw America go from a nation of small towns and farms to an urbanized, industrial society in which machines do nearly all of the heavy work (Beale 2000).

Electricity and Women

Automation during the 20th century was not limited to the farm, factory, and loading dock. Imagine a typical rural or small-town American woman living around 1900. In the morning, she might prepare breakfast on a wood-burning stove and carry buckets of water from an outside well to use for cooking and cleaning. She might devote one whole day each week to baking bread, and the preparation of a typical meal might take several hours. According to a survey conducted at the time, she would spend nearly 45 hours each week just preparing meals and cleaning up. She would spend another seven hours cleaning the house and doing laundry.

And there would have been the children to care for. Not surprisingly, most women at the turn of the century had little leisure time and little opportunity to work outside the home.

But that was about to change. After much trial and error, Thomas Edison had figured out how to produce a working electric light bulb; and by 1890, small power stations had been constructed in a number of cities and were lighting up houses for several blocks around. Edison was mostly interested in using electricity for lighting, but others found additional ways to profit from this modern miracle, and soon labor-saving devices were liberating women from much of their former drudgery. Our imaginary woman would have been delighted. By the mid-1920s, she might have owned an electric vacuum cleaner, electric washer, an iron, perhaps even an electric stove and refrigerator. Her life would have become much less burdensome, and the future of her daughters would have looked even brighter.

By 1930, 85 percent of homes in American towns and cities had electricity, and as more homes became electrified, increasing numbers of labor-saving electrical appliances became available. As a result, with each decade of the 20th century, women were able to spend less time on housework. By the mid-1920s, they were spending fewer than 30 hours per week cooking meals and cleaning up afterwards. This number continued to go down with each decade, to fewer than 20 hours per week in the 1950s and then to fewer than ten by 1975. The first counter top microwave oven was introduced in 1967. This marvelous invention, along with electric dishwashers and the growing popularity of eating out drove this number even lower. Today the typical meal prepared at home requires just over 25 minutes total for cooking and cleanup (Bowers 2000; Vanek, 1974).

In 1900, most Americans still performed hard physical work almost every day; but by 2000, most heavy labor, whether on the farm, in the factory, or in the home, had been taken over by machines. New jobs were created mostly in the service economy and few of them required physical exertion; people increasingly spent their work days sitting at a desk or behind a counter, where their "labor" consisted mostly of typing away at a computer or talking to clients or customers, either in person or on the phone. In less than a century, America became a largely sedentary nation — or a nation in which being sedentary was an option. This was a major change from our evolutionary past.

Studies show that hunter-gatherers typically walked about five to six miles daily. At an average pace of two or three miles per hour, foraging humans would spend about two to three hours walking on any given day. This typically would have consisted of short periods of activity scattered throughout the day, with numerous pauses and rests in between (O'Keefe et al. 2010). In contrast, a well-conducted study using pedometers found that modern Amish farmers walked an average of about seven to nine miles daily. This means that the typical Amish farmer in the 20th century worked harder than hunter-gatherers, spending about three to four hours of every day walking. Not surprisingly, the study did not find a single case of obesity among the Amish men studied (Roberts 2004). Of course, when it comes to American society as a whole, the Amish are extreme outliers. A recent study found that the typical American walks about 2.5 miles during the course of a day, about half as much as the average hunter-gatherer (Bassett et al. 2010).

It is easy to see how modern Americans take so few steps in the course of a day. After spending nearly all of the workday sitting at a desk, a computer work station, or in a vehicle, most of us go home to sit in front of a television, taking short reluctant walks only to the kitchen or bathroom. When we move around outside, it is almost entirely by automobile. We now buy our fast food, withdraw money from the bank, drop off or pick up dry cleaning — in fact, perform most of the chores of everyday living — without ever leaving the comfort of our home or car. According to Booth et al. (2012), "Estimated daily step numbers have declined ~ 50% to ~70% since the introduction of powered machinery." We now are beginning to see the ill effects of this all around us, and to realize that all animals, including humans, require a certain level of physical activity in order to be healthy.

A French Tale

Sometime around 1850, a 12-year-old French boy left his mountain home and walked into the broad valley below. His father, an illiterate lumberjack, had given him a silver five-franc piece and sent him away to seek his fortune, hoping his son could find a better life. The clever boy, instead of doing the easiest thing and becoming an apprentice, used this small endowment to start his own business — buying and selling feathers for pillows. He also began reading every book he could find in order to educate himself. After becoming prosperous by the age of 20, he then went into the textile business and soon became very wealthy (Mayer 1969).

Jean Mayer, grandson of the young boy who walked out of the French mountains with a five-franc piece in his pocket, was born in 1920. He grew up to become one of the world's leading authorities on food and nutrition. In a career that spanned half a century, he taught at Harvard for 25 years, wrote 750 papers and ten books, and was food policy adviser to three

American presidents. He ended his career as president of Tufts University, where he founded the first graduate school of nutrition.

He also had some interesting wartime experiences. At the outbreak of World War II, he was commissioned a second lieutenant in the French artillery. A short time later, he was captured by the Germans and sent to a prison camp. After shooting a guard and escaping, he joined the French underground and later served as an agent for British intelligence and on the staff of General Charles de Gaulle in London. He fought with the Free French and allied forces in North Africa, Italy and France, receiving a total of 14 military decorations, including the Croix de Guerre. After the war, Mayer moved to the United States, became an American citizen, earned a doctorate at Yale, and joined the faculty of Harvard University. Considering his war time exploits, it was somewhat surprising to see him described in his obituary as, "A small, bespectacled man who...never lost his French accent or Gallic jauntiness" (McFadden 1993).

Shortly after joining the Harvard faculty in 1950, Mayer conducted a study to determine how different levels of physical activity affected food consumption and body weight. Mayer and his colleagues acquired a small, motor-driven treadmill and "accustomed a large group of rats to running on it." They then divided this large group of rats into a number of smaller groups and induced them to run on the treadmill every day for varying lengths of time; for one hour, two hours, three hours, and so on. The food intake of every rat was measured and they were weighed at intervals. A control group was kept caged and was not exercised at all.

The results were revealing. The sedentary rats ate more and were heavier than those forced to take a moderate amount of exercise. This was the most significant finding of the study, showing that just a little exercise seemed to moderate appetite and limit body weight. Food intake reached its lowest point at about one hour of exercise per day. Body weight also reached a minimum with about one hour of exercise daily. Beyond that, as rats were forced to run for longer periods, they ate more and more. But their body weight stayed the same; all of the extra calories were burned up doing the extra work (Mayer et al. 1954).

This work with rats is no doubt interesting, but do humans show the same pattern? Mayer took steps to answer this question in the summer of 1954 by conducting a large-scale human study in West Bengal, a state in the eastern part of India. West Bengal is bordered by Nepal on the north and by Bangladesh on the east. Agriculture is the major economic activity and jute has long been the leading cash crop.

You might not know what jute is, but surely you have seen a burlap bag, which is made of jute. It is a fiber second only to cotton in the amount produced, and it has a wide variety of uses, including rope, twine, curtains, carpets, furniture fabric, and even clothing. Jute requires a lot of rain during the growing season and thrives on deep, wet soils in a hot climate. The West Bengal state of India, especially the southern part, has wet soils, a hot climate, and is well-watered by monsoon rains, making it ideal for cultivating jute.

Indian jute workers were good study subjects for a number of reasons. First, their jobs varied widely in the amount of physical effort required. At one extreme were sedentary clerks, whose only exercise was walking a short distance to work every day, analogous to rats kept confined to a small cage. At the other extreme were laborers tasked with lifting and carrying enormous loads of jute or swinging a heavy knife all day with little pause for rest. These workers, classified as doing very heavy work, were equivalent to the unlucky rats Mayer forced to run on a treadmill for four or more hours a day. Between these two extremes were workers doing a variety of jobs that required intermediate degrees of exertion — light, medium or heavy work.

The 300 men in the study, none more than a generation or so removed from peasant life, were uniformly short, ranging in height from just five feet, two inches to five feet, four inches. They all consumed the same traditional foods, which they bought locally and cooked themselves. As a result, it was fairly easy to keep track of how much food each man bought and consumed each week (Mayer et al. 1956).

Although not as well controlled as the rat study, the experimental conditions were about as good as one can do with human subjects outside a laboratory. The Indian workers showed the same patterns of food intake and body weight with varying degrees of exertion as did the rats. On average, the sedentary men were about 20 pounds heavier than those who were moderately active. In his published papers, Mayer presented line graphs showing how food consumption and body weight changed as activity levels increased, and the patterns for rats and humans were remarkably similar. Mayer concluded that, "in men as in rats, the subjects with regular moderate activity ate somewhat less (and were considerably thinner) than the inactive subjects" (Mayer 1968).

In both humans and rats, physical activity seemed to play a critical role in regulating appetite and weight, and it took only a small amount of activity to make a big difference. In Mayer's studies, nearly all of the benefit accrued as the test subjects of both species went from being sedentary to engaging in only moderate physical activity. Not surprisingly, no one else has done a human study as comprehensive as the one Mayer did with the jute workers. However, a number of researchers have confirmed his results in less elaborate studies (Blundell and King 1999).

Inactivity and Death

In most cases, food intake and body weight increase whenever people (or animals) adopt a sedentary existence or have one forced upon them. Over time, if given free access to rich food, sedentary animals almost always eat themselves into a state of obesity. The good news is that it takes only a small amount of habitual exercise, not too much beyond the sedentary range, for appetite and weight to be properly regulated. The bad news is that most Americans do not engage in enough physical activity needed to achieve this. If being sedentary only caused people to eat more and gain weight, it would not be such a problem, but medical experts now believe that the consequences of inactivity are much more serious than that.

English researcher Jeremiah Morris (1953) is believed to have been the first to show that being sedentary is associated with higher rates of disease and death. He and his colleagues compared the incidence of coronary heart disease among sedentary London bus drivers with that of conductors, who spent all day climbing up and down double decker busses to collect fares. The physically active conductors had a 30 percent lower incidence of coronary heart disease than drivers, whose jobs forced them to sit all day. Furthermore, conductors who were diagnosed with heart disease had less severe symptoms and were less likely to die than drivers who received a similar diagnosis.

This pioneering study was the first of many. Because of the space program, a lot of research was conducted during the second half of the 20th century to determine the effects of inactivity and weightlessness on such things as muscle mass, heart and lung function, and bone strength. In 2012, Frank W. Booth and his colleagues summarized much of this research in a long review article titled *Lack of Exercise is a Major Cause of Chronic Disease*. They based their conclusion on the "conclusive and overwhelming evidence" provided by 579 scientific papers dealing with the effects of physical inactivity on the body. Among other things, researchers had determined that prolonged inactivity results in the loss of bone and muscle mass, increased body fat, reduced insulin sensitivity, greater glucose intolerance, deterioration in heart and lung function, and accelerated aging. Additional research showed that higher levels of physical activity can extend life expectancy by as much as five to six years. The U.S. Department of Health and Human Services took note of such evidence, releasing the following statement in 2008.

> The data very strongly support an inverse association between physical activity and all-cause mortality. Active individuals — both men and women — have approximately a 30% lower risk of dying during follow-up, compared with inactive individuals.

It is gratifying to know that, after decades of research, modern researchers are rediscovering the wisdom of Hippocrates, who wrote more than 2,000 years ago,

> All parts of the body, if used in moderation and exercised in labors to which each is accustomed, become thereby healthy and well developed and age slowly; but if they are unused and left idle, they become liable to disease, defective in growth and age quickly... If we could give every individual the right amount of nourishment and exercise, not too little and not too much, we would have found the safest way to health.

We all should heed the great physician; but unfortunately, most of us will not. Instead, we will continue to eat fat, sugar, and refined carbohydrates in quantities that greatly exceed the evolutionary norms for our species. At the same time, our electrolyte intake — the amount of potassium we take in relative to sodium, will be the exact opposite of what is required for good metabolic health. Finally, most of us will not engage in the amount of physical activity our bodies need to remain healthy.

This raises an important question. Why are so many of us making such bad choices, doing things that will cause us to become unhealthy and perhaps even die prematurely? In the next chapter, we consider some of the underlying causes of food-related behavior. For example, we introduce the concept of optimal foraging, a concept that comes from a landmark 1976 paper by Paul Rozin titled *The Selection of Foods by Rats, Humans, and Other Animals*. In addition to optimal foraging, the next chapter discusses *The Sorrows of Young Werther*, a novel written more than two centuries ago by German author Johann von Goethe. This novel inspired the phrase *Werther effect*, widely used by sociologists to characterize the phenomenon of copycat suicides. We find that the ill-fated young *Werther* has something to teach us about the destructive lifestyle choices so many Americans now are making.

Chapter 11. The Omnivore's Dilemma

After more than 300 years, corn (*Zea mays*) remains the keystone species in America's food system. The typical American gets 400 to 500 kilocalories each day from just this one crop. But we eat very little of it in its natural state. Much of it comes to us in the form of beef, pork, or chicken by way of animal feed, which is used to fatten veritable cities of animals. These animal cities are called Confined Animal Feeding Operations or CAFOs. Some refer to them as Concentrated Animal Feeding Operations, fortunately with the same acronym. The thousands of animals domiciled in a CAFO eat constantly during their short lives, gain weight rapidly, and then are butchered in large numbers. Perhaps the best way to describe a CAFO is to tell the life story of one of these animals, Steer 549. The name is taken from the number on a tag attached to one of his ears.

Steer 549 was born early one spring on a ranch somewhere in central Texas. He was a big guy at birth, weighing in at about 75 pounds. He began nursing almost immediately, and within a day or so, he and his mother were set free in a nearby pasture. Steer 549 did not know who his father was. Of course, neither did his mother; the anonymous father had carried out his paternal duties by way of a semen-filled straw delivered from out of state. His semen was chosen because of its proven performance; his many offspring grew large and robust, yielding delicious fat-marbled steaks and sumptuous roasts.

Producing steaks, roasts, and hamburgers would be 549's destiny, but mercifully, he did not know that. He was happy exploring his new world and soon started supplementing his mother's milk by munching on the native grasses around him. The next months of his life were idyllic; he lived outside and roamed about as his ancestors had done for untold generations, learning to enjoy the food that evolution bred him to eat — grass. He had one really bad day when he was only a month or so old. Two humans came into the pasture, grabbed Steer 549 roughly, and then branded and castrated him. But he recovered from the pain in a short time and continued to thrive, weighing nearly 600 pounds when he was weaned at the end of September.

In only a few short months, Steer 549's grass-eating days were over. Along with many other steers, he was rounded up and hauled to a large cattle feeding yard. There he was taught to eat from a trough and his digestive system was given time to adjust to the diet he would be eating for the rest of his life. Once his system had adapted, he joined more than 35,000 other steers in an orgy of eating that lasted nearly five months. His food was prepared according to a precise recipe, scientifically formulated and mixed by computer-controlled machines. Corn was the main ingredient, supplemented by liquefied fat, concentrated protein, liquid vitamins, synthetic estrogen, antibiotics, some alfalfa hay, silage for "roughage," and even a little molasses. Huge quantities of this calorie-rich mixture were delivered to the confined animals in miles of feeding troughs. For the next five months or so, 549 and his companions had one task in life, to eat as much as possible and gain three to four pounds of body weight each day.

I doubt that 549 was aware of what was happening to his body. As one author describes it, the regimen he underwent was much like "putting a man in bed and feeding him 5,000 kilocalories a day" (Crawford and Crawford 1972). Steer 549 responded in the same way that a sedentary person fed large amounts of very palatable, energy-rich foods would. He began to grow larger and thicker, becoming at first plump and then increasingly obese. The flesh of his shoulders and flanks began to marble and he developed an imposing layer of fat around his belly.

This diet, much richer than the one he evolved to eat, also began destroying the walls of his stomach, allowing bacteria to enter his bloodstream and eventually get to his liver, where they formed abscesses. Fortunately, his human masters had things under control. They knew that 549 could not stay on this diet for too long; his liver could not take it.

So 549 did not succumb to liver failure; instead he was destined to die a quick, violent death. The reckoning came one day in late fall when he found himself trudging up a narrow, inclined chute, part of a long line of steers. He and his comrades were calm, with no idea of what awaited them a short distance ahead. Before long the chute began to descend and a bewildered 549 found himself carried along on a conveyor belt while straddling a metal bar. He could not see the "stunner," a man perched above the line of cattle on an elevated catwalk, toiling away at his bloody task. As 549 passed underneath, the unseen executioner leaned down and, with a pneumatic-powered gun, fired a long steel bolt into the forehead of the unsuspecting steer.

Reacting quickly, another worker attached a strap to one of 549's feet and secured the insensate steer to an overhead conveyor. Hanging by one back leg, 549 was carried into the bleeding area, where another busy worker quickly cut his throat. Steer 549's violent but mercifully quick death was far from unique. During a typical butchering year in the United States, perhaps three million other steers have a similar appointment with the "stunner," who surely must be the stuff of bovine nightmares.

For thousands of generations 549's ancestors ate grass — grass in the spring, grass in the summer, grass all year long, all day, every day. To get the energy needed to live, wild cattle evolved to consume two to three percent of their body weight in forage each day. This amount of food kept them healthy and well fed, but lean. Wild cattle could not afford to be obese; carrying around pounds of extra fat would slow them down and make them easy targets for predators. Survival demanded constant vigilance and, on occasion, quick action; so cattle living in the wild were never fat. In addition to being a detriment, becoming fat was not even an option. The low energy-density of their diet and the amount of activity needed to obtain it made obesity nearly impossible under natural conditions.

But the food eaten by modern feedlot cattle bears little resemblance to what their wild ancestors ate. Domesticated cattle are fed unlimited quantities of rations that are much more energy dense than grass, yielding several times more calories per unit weight or volume. According to the website of Harris Ranch, a large cattle feeding company in California, "Feedlot diets are usually very dense in food energy, to encourage the deposition of fat, or marbling, in the animal's muscles; this fat is desirable as it leads to 'juiciness' in the resulting meat. The animal may gain an additional 400 pounds...during its 3 to 4 months in the feedlot." In addition to being fed a rich diet, the animal does not have to expend any energy to acquire it; he can simply stand there and eat all day long. He has no idea how much energy he is taking in, so his body tells him to keep eating.

This regimen has very predictable and very rewarding results, at least from the cattle feeder's point of view. In a matter of months the steer fattens up and becomes marbled with fat, ready for his appointment at the slaughterhouse. It is amazing that feedlot operators have such a clear understanding of what makes animals fat, while human obesity experts, despite spending vast sums to study the problem, remain perplexed over the same process in people. This is especially ironic because the underlying process is the same in both cases.

His feeders were able to fatten Steer 549 rapidly by feeding him an energy-dense industrial diet for which evolution had not prepared him. Our "feeders" are doing much the same to us. Viewed in this way, modern humans have much more in common with feedlot animals than we might like to admit. It requires only a little imagination to think of modern cities as sprawling, concentrated animal feeding operations for people. In an animal feedlot, steers are crowded into a common area and fed a diet that is much more energy dense than the grass and other forage for which their bodies are designed.

In a similar fashion, most of our food is grown, processed, and packaged somewhere else and delivered by way of grocery stores or restaurants; and as in the case of feedlot cattle, most of that food is much more energy dense than our bodies are designed for. But humans have an advantage; unlike cattle confined in a feedlot, we have free will; we can decide what to eat and what not to eat. Unfortunately, that decision is not a simple one in today's world.

Foraging Strategies

In the latter part of the 20th century, people living in America and other industrialized nations began to have access to a mind-boggling variety of food products. Never before in history had so many been so materially blessed, but we are increasingly seeing a dark side to this abundance. We are confronted with a seemingly unstoppable epidemic of metabolic disease, reflecting our failure to solve a problem that our species has confronted many times in its history, the "omnivore's dilemma." This phrase comes from a 1976 paper by research psychologist Paul Rozin, in which he outlined a straightforward problem; if an animal species can eat almost anything, how does it decide what to eat and what not to eat? Specialty eaters such as koala bears, pandas, or lions do not face this issue. But humans can eat almost anything that walks, crawls, swims, flies, or simply grows out of the ground, so we must make choices about which foods to eat and which to avoid.

Our foraging ancestors undoubtedly used multiple strategies to determine which foods they should seek out and which ones were not worth the trouble or might even be harmful. They probably watched other animals to see what kinds of foods they favored. If monkeys and birds ate a certain kind of fruit and suffered no ill effects, chances are it was safe for humans. If wild pigs sought out a certain kind of mushroom but shunned others, perhaps humans should do the same. They also used trial and error; sampling a small amount of a new food to see how it tasted, then waiting to see if it made them sick. One can imagine a group of human foragers living in the same area for many generations. Before long, they would have catalogued all possible sources of food in their domain — every fish, bird, insect, shellfish, wild grain, nut, and mushroom. They would understand the life cycle of every plant, where they were most abundant, which parts were edible and when they were in season. This collective knowledge, so critical to survival and wellbeing, would be carefully taught to each succeeding generation.

It is easy to see how the omnivore's dilemma impacted our hunter-gatherer ancestors as they spread out across the world and encountered strange new foods, but what about today? We now buy nearly everything we eat in restaurants or food stores, so hasn't the problem gone away? If a food is there for us to purchase, can't we assume that it is good for us? The answer, of course, is no. Food marketers are in business to make as much money as possible, not necessarily to keep people healthy; and unfortunately, the unhealthiest foods are often the most profitable.

In the short term, the foods you and I buy in a supermarket or fast food restaurant seem to be safe, but long term is a different story. There is convincing evidence that eating industrial foods made from sugar, refined carbohydrates, and highly-processed vegetable oils on a regular basis eventually leads to obesity, diabetes, heart disease, and other diseases of civilization. Some researchers and health officials advise us to severely limit our consumption of these foods or even avoid them altogether. But experts on the other side, including many business and government representatives, scoff at such fears. "These are all good foods," they say. "You just have to eat them in moderation."

It is clear that the omnivore's dilemma has not gone away; instead, making healthy food choices is more difficult now than it has ever been. Consider the sheer complexity of the nutritional ecosystem in which we now live. USDA's national food composition database now contains nutrient analyses of more than 7,000 food products. You will note that I refer to items in the database as "food products," not foods, for that is what many if not most of them are, industrial products that come from a factory assembly line, concocted mostly from corn, corn syrup, sugar, refined flour, artificial flavorings, vegetable oils, and artificial preservatives. In addition to becoming harder to solve, the very nature of the omnivore's dilemma has changed.

Our foraging ancestors were concerned mostly with seeking out foods that contained a lot of fat or were very sweet. This time-tested strategy optimized nutrition for thousands of generations, but it no longer works. The modern challenge is to eat foods that assuage hunger and nourish the body without taking in such huge amounts of fat, sugar, and refined carbohydrates that obesity and other metabolic diseases become unavoidable. Unfortunately, most people have no conception of how drastically our food system changed during the 20th century and what these changes mean for our health.

The food industry spends millions of dollars each year persuading us to buy their factory-made foodstuffs, but they really do not have to work very hard at it. Although we no longer live as hunter-gatherers, we still have a strong instinct to get the most food energy for the least effort. Our foraging ancestors had the same instinct, but it served them much better. On any given day, they could spend their time in a number of ways, such as picking berries, hunting wild pigs, gathering shellfish, harvesting nuts from a nearby grove of trees, or pilfering eggs from nesting birds. It was advantageous for them to focus their efforts on those food-gathering activities that would yield the most return at the end of the day; that is, the most calories for the least effort.

Of the hundred or so foods that could be eaten, what were the ten, 15, or perhaps 20 that would yield the most food energy for the least expenditure of time and effort? This problem, though complex, was eminently solvable through trial and error; early humans probably required only a generation or so in a new area to arrive at what some ecologists call an "optimal foraging strategy" (Rozin 1976). Our species learned very early to concentrate their efforts on those natural foods that were the most energy-rich, seeking out those that were sweetest or the most fat-laden. In the wild, such foods also were the most health promoting, since, in addition to a lot of calories, they provided protein, essential fats, vitamins, and minerals; and under natural conditions it was virtually impossible to consume too much of them.

Although modern life masks this reality, we still face the same challenge — arriving at an optimal foraging and eating strategy, but the rules have changed because the human food environment has been radically altered. Sadly, the old ways will now get us in trouble. An instinctual foraging strategy that worked well throughout most of our history has become a problem in the age of industrial food. Although we no longer forage in the wild, this instinct still prompts many of us to seek out the cheapest, most calorie-rich foods available. In the 20th century, especially during the last half of it, the food industry began to enhance the sugar and fat content of foods and to make them more convenient, while at the same time reducing the cost. Modern humans, with all of their hunter-gatherer instincts still intact, were suddenly confronted with an enticing array of palate-pleasing, high-energy foods.

The response was predictable; being good hunter-gatherers, most of us followed our instincts and sought out calories in their tastiest, most convenient form. We drove our cars to fast food restaurants and supermarkets, where we eagerly purchased the cheap, calorie-dense foods filled with sugar and fat that had been so coveted throughout most of our history. As industrial foods became sweeter, fatter, and ever more convenient, we responded by consuming more and more of them. Optimal foraging strategy worked its magic, and the energy density of our diets increased far above the evolutionary norm in only a few decades — and epidemic levels of obesity and diabetes followed closely behind.

Observing the Wild Omnivore

Understanding the concept of optimal foraging made trips to the grocery store much more interesting. I began to observe the foraging habits of my fellow shoppers, noting the items they selected from the shelves and placed in their carts, and before long, definite patterns began to emerge. Modern food stores are stocked with thousands of items. Because of this, it might seem that shoppers are confronted with such a bewildering array of choices that detecting an underlying trend or foraging strategy would be very difficult. But that is not the case, because in essence, food shoppers have just three options from which to choose. Do they want their diet to consist mostly of traditional hunter-gatherer foods, traditional peasant foods, or modern industrial foods?

The hunter-gatherer option features foods such as meat, fish, leafy vegetables, berries, mushrooms, and nuts. The peasant option features grains such as corn, rice, oats, and wheat, and products traditionally made from them, such as pasta, tortillas, and bread. Legumes and pulses, such as beans, peas, and lentils are another main category of peasant food. Nearly every peasant society has depended on some combination of grains and legumes; when eaten together, they provide all of the amino acids the human body needs to make protein. Most stores sell a large number of traditional peasant foods, including many varieties of beans, peas, and lentils, either dried or canned; bags of rice, flour, and corn meal; and shelves of pasta products.

In addition to traditional hunter-gatherer and peasant foods, modern food stores devote a lot of shelf space to highly-processed, industrial foods. Most of these foods are formulated in factories and consist of varying combinations of highly-refined carbohydrates, high-fructose corn syrup, salt, cheap vegetable oils, and artificial coloring and flavoring agents. Included are such things as cookies; candy; chips; sugary cereals; frozen, pre-cooked meals; and sweetened drinks.

The fact that the three main categories of human food tend to occupy different areas of the store made it rather easy to see different shopping patterns. People who were lean or of normal weight (the lean phenotypes) tended to shop mostly in the hunter-gatherer and peasant food sections, choosing things such as nuts, fruits, vegetables, whole grains, milk, fish, and fresh meat, while selecting only limited numbers of highly-processed industrial foods. The overweight or obese phenotypes, almost without exception, did exactly the opposite. They shopped mostly in the aisles devoted to industrial foods, filling their carts with items such as potato chips, cookies, ice cream, sugar-infused breakfast cereals, pre-packaged dinners, and soft drinks.

It is easy to see why so many people are drawn to industrial foods. First, such foods are very convenient. Many of them can be eaten directly out of a bag or box, which then can simply be thrown away. Even those that require cooking, such as frozen entrees or frozen dinners, usually come in microwaveable containers that can be disposed of easily after the food is eaten. In addition to ease of preparation and cleanup, mass-produced industrial foods generally are very cheap on a per calorie basis. The refined carbohydrates (usually from corn, wheat, or rice), high fructose corn syrup, and vegetable oils used to make them are heavily subsidized by the U.S. Department of Agriculture, keeping costs for food manufacturers very low. In addition, mass production helps drive the cost per calorie even lower.

Peasant foods, such as flour, cornmeal, pasta, rice, lentils, and beans are relatively inexpensive, but it takes a lot of time to cook them and to clean up after the meal. Traditional hunter-gatherer foods, such as meats, fish, and fresh vegetables, also require more preparation and cleanup time. In addition, they are much more expensive than either industrial or peasant foods. Industrial foods are abundant, relatively cheap, heavily subsidized, and heavily promoted. Unfortunately, they are far more energy dense than the foods to which humans are metabolically adapted. They also contain levels of refined carbohydrates, sugar, and sodium far in excess of anything evolution has prepared us for.

For most of human evolution, selecting healthy foods was an easy thing to do, but that is no longer true. We now live in a daunting food environment, one in which choosing the right foods requires careful thinking and the exercise of will. Anyone wishing to avoid obesity, diabetes, strokes, heart attacks, and perhaps even cancer, must consciously limit their consumption of mass-produced industrial foods and instead eat traditional hunter-gatherer or peasant foods. We no longer can rely on an instinctive foraging strategy, which tells us to eat whatever provides the most calories and is easiest to acquire. Nor can we simply watch those around us to see what they are eating. That worked throughout most of human history, but doing so can get you into trouble in today's world.

The Sorrows of Young Werther

In addition to ignoring our innate foraging instincts, eating well and staying healthy in the modern food environment requires that we resist some powerful social conditioning. A literary tale from the late 1700s will help to explain this. In 1774, German author Johann von Goethe published *The Sorrows of Young Werther*, a book about a sensitive young artist who commits suicide. This book made Goethe into perhaps the world's first international literary celebrity. It also sparked a wave of copycat suicides; as the novel gained in readership, young people all over Europe started killing themselves, with the total number of dead believed to have exceeded 2,000.

It might surprise you to know that this is not a unique or even rare phenomenon; imitative suicide, now referred to as the *Werther effect*, is disturbingly common. One researcher analyzed suicide statistics in the United States from 1947 to 1968 and found that during the two months following the appearance of a suicide story on the front page of a major newspaper, an average of 58 more people than usual took their lives (Phillips 1974). The suicide rate would then level off again until the next major story appeared. Subsequent research has shown that television news stories are equally effective in causing imitative suicide (Bollen and Phillips 1982).

Imitative suicide is a striking example of a widespread human tendency. In almost every area of life, we instinctively look to those around us when uncertain about what to do. As in the case of those mimicking Goethe's tragic young hero, imitative behavior can sometimes result in great harm.

But in general the instinct to behave in the same way as those around us is beneficial and probably originated very early in our evolution. Experience taught our ancestors, even before they became human, that, when in doubt, the best course of action was to follow the crowd.

Modern food marketers know this natural instinct well; they understand how to capitalize on our innate compulsion to do as others do. If a restaurant chain can get the message across that everyone is flocking there to eat (90 Billion Hamburgers Sold!), they are well on their way to success. A cereal producer can become rich by convincing children that all of their peers, especially the cool kids, are clamoring to eat his particular brand of cereal, favoring it above all others. It is no surprise that two of the most common phrases used in advertising are "fastest-growing" and "largest-selling." The not-so-hidden message: the rest of the tribe is eating this, so you should be eating it too. One should not underestimate the power of the *Werther effect*. If it can induce people to kill themselves, it also can exert a profound influence on what we eat. The *Werther effect*, or imitative suicide, is not an inappropriate label for what is happening in modern America. Day by day, meal by meal, all too many of us are committing dietary suicide.

Epilogue. A Deadly Inheritance

When it comes to food and health, the next century will be a tragic one for many Americans. Agricultural science and technology — along with a very large portion of the world's best farmland, have given our nation everything needed to provide every American with a healthy, nourishing diet. Unfortunately, that promise remains unfulfilled. In addition to being divided by race, ethnicity, gender, and socioeconomic status, we increasingly are being divided into two nations nutritionally, two distinct populations living in very different food environments, with different health outcomes and different life expectancies.

An earlier chapter featured a discussion of some former American presidents and how long they lived. Let us return to that topic by considering the longevity of the last five presidents to serve during the 20th century, from Gerald Ford, who became president in 1974 and served until 1977 to Bill Clinton, who was president from 1993 to 2001. Below is a list of these five men, showing the ages at which each died, or in the cases of Presidents Clinton and Carter, their current ages.

Gerald Ford — Died at age 93
Jimmy Carter — Alive at age 97
Ronald Reagan — Died at age 93
George H.W. Bush — Died at age 94
Bill Clinton — Alive at age 75

The trend is obvious. American presidents now are living to a very old age. Why is this happening and what does it have to say about the future prospects of Americans at large? I gained some insight into this question a few years ago while visiting a sick family member at the Duke University Medical Center in Durham, North Carolina. For about two weeks, I spent part of nearly every day at this large teaching hospital that, in addition to attracting people of great wealth, provides care to many poor people from throughout the region.

On most days I would arrive early, leave my car in a nearby parking garage, and walk through a long tunnel to the hospital. Some mornings I would get a cup of coffee and sit at the hospital end of the tunnel, watching doctors and medical students arriving to begin their workday. At other times of the day I would observe while walking the halls or while sitting in the hospital cafeteria.

I was struck by the fact that nearly every person I saw wearing a white coat with a stethoscope draped over his or her shoulder was of normal weight. Although I did not make a precise count, fewer than five percent (more likely two to three percent) of the medical professionals showed any evidence of obesity. It is likely that they also had low rates of diabetes, hypertension, and other types of metabolic disease. This was in sharp contrast to what I saw in the crowded waiting areas of the various treatment clinics. In many cases, well over half of the people sitting there were obese and most of the remainder were overweight. A dismaying number had undergone limb amputations.

The striking contrast between patients and doctors at the Duke University Medical Center provides valuable insight into who will be obese and sick and who will not in 21st century America. Medical training versus no medical training is not the issue here. The important thing is that doctors and medical students at a prestigious, selective institution such as Duke represent a particular demographic, a competitively selected group of people who are very intelligent, well-educated, and affluent. Of course, there are intelligent, well-educated, and affluent people among the patients also, but the patient group as a whole more closely resembles mainstream America, so the overall average is lower when it comes to intelligence, education, and income; and these are the three factors that will increasingly divide America into the obese, diabetic majority and a small minority that does not become obese and remains largely free of diabetes and early heart disease.

It is easy to predict what our nation will look like in the year 2050. In a word, America will look fat. Only about two out of ten adults will be of normal weight — that is, according to today's standards; by 2050, being overweight or perhaps even obese will have become the new normal. One third or more of Americans will be diabetic, with the percentage afflicted in some minority populations exceeding 50 percent. We truly will be a nation divided, comprised of two distinct subpopulations occupying the same geographic area but living vastly different lives.

This is hardly the picture of a rational, humane society, but there is little chance of things getting better soon. For the foreseeable future, clever, innovative people will continue to get rich manufacturing and selling food products that in time make people obese and sick. Then, adding to the injustice, other clever people will make a good living treating these victims of diet-induced illness. A food and "health care" system intertwined in this way makes economic sense, but only in the most twisted Machiavellian sort of way; in terms of basic human values, it represents folly and dysfunction on a grand scale. When it comes to food and nutrition in America, social Darwinism clearly prevails. We are seeing the results of a vast, unplanned experiment; feeding unlimited amounts of highly-processed, energy-dense industrial foods to a large population of humans, most of whom do not get the minimal amount of physical activity needed to maintain metabolic health.

It is impossible to predict the long-term outcome with certainty, but considering what is already happening, it is likely to be very grim. Health officials will continue to express dismay at the devastating effects of metabolic disease on human health, appearance, and quality of life. America will continue to spend millions of dollars every year studying this problem and even more in largely useless attempts to solve it. The level of alarm will keep increasing, paralleling the rapid rise in the number of people afflicted. Theoretically, the problem is solvable. We could simply dismantle our current industrial system of agriculture and food production and replace it with one designed to produce and deliver traditional foods; but that is very unlikely to happen. Our population has grown too large and there are too many people with vested interests in the present system. As a society at large, there is no practical way to turn back from our present course.

Instead, things are likely to become even worse, because humans, unlike feedlot animals, produce offspring, and therein lies a problem. It appears that metabolic disease can be passed directly from one generation to the next. Mothers who suffer from obesity and diabetes tend to have larger and fatter babies. If the mother has a lot of sugar and insulin in her blood, the same will be true of the developing child. In response to high amounts of glucose in the mother's blood, the child's pancreas can develop more insulin-secreting cells. The higher the blood sugar of the pregnant mother, the more insulin-secreting cells the child will tend to have. The baby will be born with more fat cells and a tendency to oversecrete insulin, primed to become obese and diabetic at an early age. If current trends continue, the number of such children will increase with each generation, and they will become obese and diabetic at increasingly younger ages. With each generation, more and more children will be fattened and sickened in the womb, and this does not bode well for the future.

One is reminded of Exodus 34:7, which warns of a disapproving God "visiting the iniquity of fathers on the children and on the grandchildren to the third and fourth generations." This biblical text could well be a description of modern America. We are a rich but benighted nation caught in a downward spiral of sickness and early death. Birth by birth, generation by generation, we are drifting ever downward to an uncertain end; and sadly, we have only ourselves to blame. As the comic strip character Pogo famously proclaimed, "We have met the enemy and he is us."

References

Armstrong, G.L., Conn, L.A. and Pinner, R.W. 1999. Trends in infectious disease mortality in the United States during the 20th century. *JAMA* 281(1): 61-66.

Armstrong, B., Van Merwyk, A.J. and Coates, H. 1977. Blood pressure in Seventh Day Adventists vegetarians. *American Journal of Epidemiology* 105: 444-449.

Arnot, R. 1997. *Dr. Bob Arnot's revolutionary weight control program.* Little, Brown and Company, Boston, MA.

Atkins, R.C. 1972. *Dr. Atkins' diet revolution: The high calorie way to stay thin forever.* David McKay Company, Philadelphia, PA.

Banting, W. 1863. *Letter on corpulence: Addressed to the public.* London. Reprinted 2005. Cosimo Classics, New York.

Beale, C.L. 2000. A century of population growth and change. *Food Review* 23(1): 16-22.

Benjamin, E.J., Muntner, P., Alonso, A. et al. 2019. Update: A report from the American Heart Association. *Circulation* 139(10): e56-e66.

Bennett, P. 1999. Type 2 diabetes among the Pima Indians of Arizona: an epidemic attributable to environmental change? *Nutrition Reviews* 57(5): S51-S54.

Bennett, W. 1995. Beyond overeating. *New England Journal of Medicine* 332: 673-74.

Bennett, W. and J. Gurin. 1982. *The dieter's dilemma.* Basic Books, New York.

Blakeslee, A. and Stamler, J. 1966. *Your heart has nine lives: Nine steps to heart health.* Pocket Books, New York.

Blundell, J.E. and King, N.A. 1999. Physical activity and regulation of food intake: current evidence. *Medicine and Science in Sports and Exercise* 31(11): S573-S583.

Bollen, K.A. and Phillips, D.P. 1982. Imitative suicides: A national study of the effects of television news stories. *American Sociological Review* 47: 802-809.

Bondi, A. 1982. Nutrition and animal productivity. In Recheigl, M. (Ed). *CRC Handbook of Animal Productivity*. CRC Press, Boca Raton, FL.

Booth, F.W., Roberts, C.K. and Laye, M.J. 2012. Lack of exercise is a major cause of chronic disease. *Comprehensive Physiology* 2(2): 1143-1211.

Bowers, D.E. 2000. Cooking trends echo changing roles of women. *Food Review* 23(1): 23-29. Economics Research Service, USDA, Washington, DC.

Braidwood, R.J. 1957. *Prehistoric men*. Chicago Natural History Museum Popular Series, Anthropology, No. 37, Chicago, IL.

Brownell, K.D. and Horgen, K.B. 2004. *Food fight: The inside story of the food industry, America's obesity crisis, and what we can do about it*. Contemporary Books, Chicago, IL.

Budiansky, S. 1999. *The covenant of the wild: Why animals chose domestication*. Yale University Press, New Haven, CT.

Carpenter, W.H., Fonong, T., Toth, M.J. et al. 1998. Total daily energy expenditure in free-living older African-Americans and Caucasians. *American Journal of Physiology* 274 (Part 1): E96-E101.

Carr, G. 2005. The proper study of mankind. *The Economist* 377 (8458): 3-4.

Clarke, F.W. (1924). *Data of geochemistry*. United States Geological Survey, Reston, VA.

Cohen, M. 1977. *The Food crisis in prehistory: Overpopulation and the origins of agriculture*. Yale University Press, New Haven, CT.

Cordain, L., Miller, J.B., Eaton, S.B. et al. 2000. Plant-animal subsistence ratios and macronutrient energy estimations in worldwide hunter-gatherer diets. *American Journal of Clinical Nutrition* 71(3): 682–92.

Colles, L. 1999. *Fat: exploding the myths*. Welcome Rain Publishers, New York.

Cook, J. 1784. *A voyage to the Pacific Ocean*. G. Nichol, London.

Coppinger, R. and Coppinger, L. 2001. *Dogs: A startling new understanding of canine origin, behaviour and evolution*. Scribner, New York.

Cossrow, N. and Falkner, B. 2004. Race/ethnic issues in obesity and obesity-related comorbidities. *Journal of Clinical Endocrinology and Metabolism* 89(6): 2590-2594.

Crawford , M. and S. Crawford. 1972. *What we eat today*. Spearman, London.

Crookes, W. 1899. *The wheat problem*. Chemical News Office, London.

Dabalea, D. 2007. The predisposition to obesity and diabetes in offspring of diabetic mothers. *Diabetes Care* 30 (Supplement 2): S169–S174.

Darwin, C. 1871. *The descent of man*. John Murray, London.

Diamond, J. 1987. The worst mistake in the history of the human race. *Discover*: 64-66.

Dies, E.J. 1949. *Titans of the soil*. University of North Carolina Press, Chapel Hill, NC.

DiNicolantonio, J.J., Lucan, S.C. and O'Keefe, J.H. 2016. The evidence for saturated fat and for sugar related to coronary heart disease. *Progress in Cardiovascular Diseases* 58(5): 464-472.

Donaldson, B.F. 1961. *Strong medicine*. Cassell, New York.

Dwyer-Lindgren, L., Mackenbach, J.P., Lenthe van, F.J. et al. 2016. Diagnosed and undiagnosed diabetes by county in the U.S., 1999-2012. *Diabetes Care* 39(9): 1556-1562.

Dye, T. and D.W. Steadman. 1990. Polynesian ancestors and their animal world. *American Science* 78(3): 207-215.

Eaton, S.B. and M. Konner. 1985. Paleolithic nutrition: a consideration of its nature and current implications. *New England Journal of Medicine* 312: 283-289.

Edholm, O.G., Fletcher, J.M., Widdowson, E.M. and McCance, R.A. 1955. The energy expenditure and food intake of individual men. *British Journal of Nutrition* 9: 286-300.

Eyre, E.J. 1845. *Journals of expeditions of discovery into central Australia, and overland from Adelaide to Prince George's Sound, in the years 1840-41*. Boone, London.

Farb, P. 1978. *Humankind*. Houghton Mifflin Company. Boston.

Farb, P. 1968. *Man's rise to civilization as shown by the Indians of North America from primeval times to the coming of the industrial state*. Avon Books, New York.

Farb, P. and Armelagos, G. 1980. *Consuming passions: The anthropology of eating*. Houghton Mifflin Co., Boston, MA.

Fumento, M. 1997. *The fat of the land*. Penguin Putnam, New York.

Gade, D.W. 2000. Hogs (pigs). In Kiple, K. F. and Kriemhild, C.O. (Ed.). 2000. *The Cambridge world history of food*. Cambridge University Press, Cambridge, UK.

Gaesser, G.A. 1996. *Big fat lies*. Fawcett Columbine, New York.

Galton, L. 1976. *The truth about fiber in your food*. Crown Publishers, New York.

Gannon, B., DiPietro, L. and Poehlman, E.T. 2000. Do African Americans have lower energy expenditure than whites? *International Journal of Obesity Related Metabolic Disorders* 24(1): 4-13.

Gates, P. 1960. *The farmer's Age: Agriculture: Economic history of the United States. vol. 3*. Holt, Rinehart and Winston, New York.

Geissler, C.A. and M.S. Aldouri. 1985. Racial differences in the energy cost of standardized activities. *Annals of Nutrition and Metabolism* 29(1): 40-47.

Gerrior, S. and Bente, L. 1997. *The nutrient content of the U.S. food supply 1909-94*. Home Economics Research Report 53. U.S. Department of Agriculture, Washington, DC.

Gibbs, J.R.C., Young, R.C., and Smith, G.P. 1973. Cholecystokinin decreases food intake in rats. *Journal of Comparative and Physiological Psychology* 84: 488-495.

Glick, Z. and J. Mayer.1968. Preliminary observations on the effect of intestinal mucosa extract on food intake of rats. *Federation Proceedings* 27: 485.

Goran, M.I. and Weinsier, R.L. 2000. Role of environmental vs. metabolic factors in the etiology of obesity: Time to focus on the environment. *Obesity Research* 8(5): 407-409.

Grady, D. 2008 (January 23). Diabetes study favors surgery to treat obese. *New York Times online*.

Gregg, E.W., Zhou, X., Cheng, Y.J. et al. 2014. Trends in lifetime risk and years of life lost due to diabetes in the USA, 1895 to 2011: A modeling study. *Lancet Diabetes & Endocrinology* 2(11): 867-874.

Grey, G. 1841. *Journal of two expeditions of discovery in north-west and western Australia, during the years 1837, 38, and 39*. Boone, London.

Griffin, J.S. 1943. *A doctor comes to California: The diary of John S. Griffin, assistant surgeon with Kearny's dragoons, 1846-1847*. California Historical Society, San Francisco, CA.

Groen, J.J., Tijong, K.B., Koster, M. et al. 1962. The influence of nutrition and ways of life on cholesterol and the prevalence of hypertension and heart disease among Trappist and Benedictine monks. *American Journal of Clinical Nutrition* 10: 456-470.

Hales, C.M., Fryar, C.D., Carroll, M.D. et al. Trends in obesity and severe obesity prevalence in US youth and adults by sex and age, 2007-2008 to 2015-2016. JAMA 319(16): 1723-1725.

Hall, R.H. 1976. *Food for naught: The decline in nutrition*. Vintage Books, New York.

Harper, A.E. 1969. Where are we? Where are we going? *American Journal of Clinical Nutrition* 22: 87-98.

Harris, J. and Benedict, F. 1918. A biometric study of human based metabolism. *Proceedings of the Natural Academy of Sciences* 4(12): 370-373.

Harris, M. 1998. *Good to eat: Riddles of food and culture.* Waveland Press, Prospect Heights, IL.

Hathaway, M.L. 1959. Trends in heights and weights. In Stefferud, A. (Ed). *Food: The yearbook of agriculture 1959.* U.S. Government Printing Office, Washington, DC.

Hellmich, N. 2010 (November 8). Fast food survey: Nearly all kids' meals high in salt, calories. *USA Today.*

Hirsch, J. and Gallian, E. 1968. Methods for the determination of adipose cell size in man and animals. *Journal of Lipid Research* 9: 110-119.

Howard, D.A. 1997. *Conquistador in chains: Cabeza de Vaca and the Indians of the Americas.* University of Alabama Press, Tuscaloosa, AL.

Hudson, B.D. 1992. The soil survey as paradigm-based science. *Soil Science Society of America Journal* 56(3): 836-841.

Jakicic, J.M. and Wing, R.R. 1998. Differences in resting energy expenditure in African-American vs Caucasian overweight females. *International Journal of Obesity Related Metabolic Disorders* 22(3): 236-242.

Jones. D.S., Podolsky, S.H and Greene, J.A. 2012. The burden of disease and the changing task of medicine. *New England Journal of Medicine* 366(25): 2333-2338.

Joslin, E.P. 1934. Studies in diabetes mellitus. II: Its incidence and the factors underlying its variations. *American Journal of Medical Science* 187(4): 433-457.

Kannel, W.B., Dawber, T.R., Kagan, A. et al. 1961. Factors of risk in the development of coronary heart disease – Six-year follow-up experience. *Annals of Internal Medicine* 55(1): 33-50.

Karlen, A. 1995. *Man and microbes: Disease and plagues in history and modern times.* G.P. Putnam's Sons, New York.

Karppanen, H. and Mervaala, E. 1996. Adherence to and population impact of nonpharmacological and pharmacological antihypertensive therapy. *Journal of Human Hypertension* 10 (Supplement 1): S57-S61.

Keys, A. 1980. *Seven countries: A multivariate analysis of death and coronary heart disease.* Harvard University Press, Cambridge, MA.

Keys, A. 1956. The diet and development of coronary heart disease. *Journal of Chronic Disease* 4(4): 364-380.

Keys, A. 1952. Human atherosclerosis and the diet. *Circulation* 5(1): 115-118.

Keys, A. and Anderson, J.T. 1954. The relationship of the diet to the development of atherosclerosis in man. *Symposium on Atherosclerosis*. National Academy of Science-National Research Council, Washington, DC.

Keys, A., Brozek, J., Henschel, A., Mickelsen, O. and Taylor, H.L. 1950. *The biology of human starvation*. University of Minnesota Press, Minneapolis, MN.

Kiple, K.F. 2007. *A moveable feast: Ten millennia of food globalization*. Cambridge University Press, New York.

Kleinfeld, N.R. 2006 (January 9). Diabetes and its awful toll quietly emerges as a crisis. *New York Times online*.

Klemmer, P., Grim, C.E. and Luft, A.C. 2014. Who and what drove Walter Kempner: the rice diet revisited. *American Heart Journal* 64: 684.

Kojima, M., Hosoda, H., Date, Y., Nakazato, M., Matsuo, H. and Kangawa, K. 1999. Ghrelin is a growth-hormone-releasing acylated peptide from stomach. *Nature* 402: 656-660.

Kolata, G. 1998 (December 29). Scientist at work: Dean Ornish; promoter of programs to foster heart health. *New York Times online*, www.nytimes.com. Accessed March 11, 2021.

Koma, J.W., Vercammen, K.A., Jarienski, M.P. et al. 2019. Sugary drink consumption by supplemental nutritional assistance program status. *American Journal of Preventive Medicine* 58(1): 69-78.

Kosaka, T. and Lim, R. 1930. Demonstration of the humoral agent in fat inhibition of gastric secretion. *Proceedings of the Society for Experimental Biology and Medicine* 27: 890-891.

Kreuger, A. 2005 (January/February). How'd I get so fat? *AARP Magazine*: 48-54, 81.

Lee, R.B. 1998. What hunters do for a living, or, how to make out on scarce resources. In Gowdy, J. (Ed.). *Limited wants, unlimited means: A reader on hunter-gatherer economics and the environment*. Island Press, Washington, DC.

Lee, R.B. and Devore, I. (Ed). *Man the hunter*. Aldine, Chicago, IL.

Legge, A.J. and Rowley-Conwy, P.A. 1987. Gazelle killing in stone age Syria. *Scientific American* 257: 88-95.

Lieb, C.W. 1929. The effects on human beings of a twelve months' exclusive meat diet based on intensive clinical and laboratory studies on two arctic explorers living under average conditions in a New York climate. *JAMA* 93(1): 20-22.

Leibel, R., Rosenbaum, M. and Hirsch, J. 1995. Changes in energy expenditure resulting from altered body weight. *New England Journal of Medicine* 332: 626.

MacCleery, D.W. 1992. *American forests: A history of resilience and recovery*. USDA Forest Service and Forest History Society, Durham, NC.

MacLagan, N.F. 1937. The role of appetite in the control of body weight. *Journal of Physiology* 90: 385-394.

Mann, C. C. 2002. 1491. *Atlantic Monthly*: 4154.

Mann, C.C. 2005. Squanto and the pilgrims: native intelligence. *Smithsonian* 36(9): 95-108.

Mann, G.V., Shaffer, R.D., Anderson, R.S. et al. 1964. Cardiovascular disease in the Masai. *Journal of Atherosclerosis Research* 4(4): 289-312.

Mann, G.V., Spoerry, A., Gary, M. and Jarashow, D. 1972. Atherosclerosis in the Masai. *American Journal of Epidemiology* 95(1): 26-37.

Mayer, J. 1969. Andre Mayer – A biographical sketch. *Journal of Nutrition* 99: 1-8.

Mayer, J. 1968. *Overweight: causes, cost and control.* Prentice-Hall, Englewood Cliffs, NJ.

Mayer, J., Marshall, N.B., Vitale, S.S. et al. 1954. Exercise, food intake and body weight in normal rats and genetically obese adult mice. *American Journal of Physiology* 177: 544-548.

Mayer, J., Roy, P. and Mitra, K.P. 1956. Relation between caloric intake, body weight, and physical work. *American Journal of Clinical Nutrition* 4: 169-175

Menke, A., Casagrande, S., Geiss, L. and Cowie, C.C. 2015. Prevalence of and trends in diabetes among adults in the United States, 1988-2012. *JAMA* 314(10): 1021-1029.

McCarthy, F.D. and McArthur, M. 1960. The food quest and the time factor in aboriginal economic life. In Mountford, C.P. (Ed.). *Records of the Australian-American scientific expedition to Arnhem Land. Volume 2: Anthropology and nutrition.* Melbourne University Press, Melbourne.

McNeill, W.H. 1998. *Plagues and peoples.* Anchor Books/Doubleday, New York.

Melville, H. 1846. *A narrative of four months residence among the natives of a valley of the Marquesas Islands.* John Murray, London.

Mintz, S.W. 1985. *Sweetness and power: The place of sugar in modern history.* Penguin, New York.

Moore, R.D. 2001. *The high blood pressure solution: A scientifically proven program for preventing strokes and heart disease.* Healing Arts Press, Rochester, VT.

Morris, J.N., Heady, J.A., Raffle, P.A. et al. 1953. Coronary heart disease and physical activity of work. *Lancet* 265: 1053-1057.

Moss, M. 2010 (May 30). The hard sell on salt. *New York Times.*

Mount, J.L. 1975. *The food and health of western man.* John Wiley and Sons, New York.

Mrosovsky, N. and Sherry, D.F. 1980. Animal anorexias. *Science* 207: 837-842.

Murdock, G.P. 1968. The current status of the world's hunting and gathering people. In Lee, R.B. and Devore, I. (Eds.). *Man the hunter*. Aldine, Chicago.

Narayan, K.M.V., Boyle, J.P., Thompson, T.J. et al. 2003. Lifetime risk for diabetes mellitus in the United States. *Journal of the American Medical Association* 290(14): 1884-1890

Neel, J.V. 1962. Diabetes mellitus: A "thrifty" genotype rendered detrimental by "progress"? *American Journal of Human Genetics* 14: 353-362.

Oakley, K. P. 1976. *Man the tool-maker*. University of Chicago Press, Chicago, IL.

O'Keefe, J.H., Vogel, R., Lavie, C.J., and Cordain, L. 2010. Achieving hunter-gatherer fitness in the 21st century: Back to the future. *American Journal of Medicine* 123: 1082-1086.

Olshansky, S J., Oassaro, D.J., Hershow, R.C. et al. 2005. A potential decline in life expectancy in the United States in the 21st century. *New England Journal of Medicine* 352: 1138-1145.

Ophir, O., Peer, G. Gilad, J. et al. 1983. Low blood pressure in vegetarians: The possible role of potassium. *American Journal of Clinical Nutrition* 37: 755-762.

Ornish, D. 1993. *Eat more, weigh less*. HarperCollins, New York.

Osler, W. 1892. *The principles and practice of medicine*, 4th edition. D. Appleton, New York.

Paffenberger, R. and Hale, W. 1975. Work activity and coronary heart mortality. *New England Journal of Medicine* 292: 545-550.

Page, J. 1983. Forest (Planet Earth). Time-Life Books, Chicago, IL.

Pan, L., Galuska, D.A., Sherry, B. et al. 2009. Differences in prevalence of obesity among black, white, and Hispanic adults – United States, 2006-2008. *Morbidity and Mortality Weekly Report* 58(27): 740-744.

Pasquet, P., Brigant, L., Froment, A. et al. 1992. Massive overfeeding and energy balance in men: The Guru Walla model. *American Journal of Clinical Nutrition* 56: 583-490.

Pennington, 1954. Treatment of obesity: Developments of the past 150 years. *Journal of Digestive Diseases* 21(3): 65-69.

Pennington, A. 1953. Reorientation on obesity. *New England Journal of Medicine* 248: 959-964.

Pfeiffer, J.E. 1969. *The emergence of man*. Harper & Row, New York.

Phillips, D.P. 1974. The influence of suggestion on suicide: Substantive and theoretical implications of the Werther effect. *American Sociological Review* 39: 340-354.

Pollan, M. 2006. *The omnivore's dilemma*. The Penguin Press, New York.

Pool, R. 2001. *Fat: Fighting the obesity epidemic*. Oxford University, Oxford, UK.

Porter, D. 1823. *A voyage in the south seas*. Richard Phillips and Co., London.

Postman, N. 1989 (December). Learning by story. *Atlantic Monthly*: 119-124.

Prentice, A.M., Rayco-Solon, P. and Moore, S.E. 2005. Insights from the developing world: Thrifty genotypes and thrifty phenotypes.

Pritchard, J.B (Ed.). 1969. *Ancient near eastern texts relating to the Old Testament*. Princeton University Press, Princeton, NJ.

Pyke, M. 1970. *Man and food*. McGraw-Hill Book Co., New York.

Ravnskov, U. 2000. *The cholesterol myths: Exposing the fallacy that saturated fat and cholesterol cause heart disease*. New Trends, Washington, DC.

Redding, R.W. 1988. A general explanation of subsistence change: From hunting and gathering to food production. *Journal of Anthropological Archaeology* 7: 56-97.

Reeves, A. and F. Plum. 1969. Hyperphagia, rage and dementia accompanying a ventromedial hypothalamic neoplasm. *Archives of Neurology* 20: 616-624.

Roberts, W.C. 2004. The Amish, body weight, and exercise. *American Journal of Cardiology* 94: 1221.

Root, W. and R. de Rochemont. 1976. *Eating in America*. William Morris, New York.

Rousseau, J. 1755. *The social contract and discourses*. Paris.

Rozin, P. 1976. The selection of foods by rats, humans, and other animals. Pp. 21-76 In Rosenblatt et al. (Ed.). *Advances in the Study of Behavior* 6. Academic Press, New York.

Rush, E.C., Plank, L.D. and Robinson, S.M. 1997. Resting metabolic rate in young Polynesian and Caucasian women. *International Journal of Obesity Related Metabolic Disorders* 21(11): 1071-1075.

Russell, F. 1975. *The Pima Indians*. University of Arizona Press, Tucson, AZ.

Sacks, F.M., Rosner, B. and Kass, E.H. 1974. Blood pressure in vegetarians. *American Journal of Epidemiology* 100(5): 390-398.

Saundby, R. 1891. *Lectures on diabetes: Including the Bradshaw lecture, delivered before the royal college of physicians on August 18th, 1890*. E.B. Treat, New York.

Sellayah, D., Cagampang, F.R. and Cox, R.D. 2014. On the evolutionary origins of obesity: A new hypothesis. *Endocrinology* 155(5): 1573-1588.

Shaper, A.G. 1962. Cardiovascular studies in the Samburu tribe of northern Kenya. *American Heart Journal* 63(4): 437-442.

Shell, E.R. 2002. *The hungry gene: The science of fat and the future of thin.* Atlantic Monthly Press, New York.

Shintani, T.T., Hughes, C.K., Beckham, S. and O'Connor, H.K. 1991. Obesity and cardiovascular risk intervention through the ad libitum feeding of traditional Hawaiian diet. *American Journal of Clinical Nutrition* 53:1647S-1651S.

Simmons, A. 2007 (August). Fat, Inc.: How the XXLing of America is becoming big business. *Readers Digest*: 104-109.

Sims, E.A. 1974. Studies in human hyperphagia. In Bray, G. and Bethune, J. (Ed.) *Treatment and management of obesity*. Harper and Row, New York.

Sjodin, A.M., Andersson, A.B., Hogberg, J.M., and Westerterp, K.R. 1994. Energy balance in cross country skiers. A study using doubly labeled water and dietary record. *Medicine and Science in Sports and Exercise* 26: 720-724.

Smil, V. 2001a. *Enriching the Earth: Fritz Haber, Carl Bosch, and the transformation of world food production*. The MIT Press, Cambridge, MA.

Smil, V. 2001b. Feeding the world: A challenge for the twenty-first century. MIT Press, Cambridge, MA.

Smith, R.E. and Fairhurst, A.S. 1958. A mechanism of cellular thermogenesis in cold-adaptation. *Proceedings of the National Academy of Sciences* 44: 705-711.

Speakman, J.R. 2008. Thrifty genes for obesity, an attractive but flawed idea, and an alternative perspective: The 'drifty gene' hypothesis. International Journal of Obesity 32: 11): 1611-1617.

Stamenov, M. and Gallese, V. (Eds.). *Mirror neurons and the evolution of brain and language*. John Benjamins Publishing, Amsterdam.

Stefansson, V. 1921. *The friendly arctic: The story of five years in polar regions*. Greenwood Press, New York.

Stefansson, V. 1946. *Not by bread alone*. The MacMillan Company, New York.

Strobel, A., Issad, T., Camoin, L. et al. 1998. A leptin missence mutation associated with hypogonadism and morbid obesity. *Nature Genetics* 18: 213-215.

Sykes, B. 2001. *The seven daughters of Eve*. Corgi Books, London.

Tanner, T.H. 1869. *The practice of medicine*. Henry Renshaw, London.

Teicholz, N. 2014. *The big fat surprise: Why butter, meat, and cheese belong in a healthy diet*. Simon & Schuster, New York.

Todhunter, E.N. 1959. The story of nutrition. In Stefferud, A. (Ed.). *The Yearbook of Agriculture 1959*. U.S. Government Printing Office, Washington, DC.

U.S. Bureau of the Census. 1975. *Historical statistics of the United States from colonial times to 1970, bicentennial edition, part 1.* U.S. Department of Commerce, Washington, DC.

USDA. 2000. A taste of the 20th century. *Food Review* 23(1): 1-2.

USDA. 2002. *Agricultural fact book, 2001-2002.* U.S. Government Printing Office, Washington, DC.

USDA. 2000. *Agriculture fact book 2000.* U.S. Government Printing Office, Washington, DC.

Vanek, J. 1974. Time spent in housework. *Scientific American* 231(5): 116-120.

Vernon, M.C., Mavropoulos, J., Transue, M. et al. 2003. Clinical experience of a carbohydrate-restricted diet: Effect on diabetes mellitus. *Metabolic Syndrome and Related Disorders* 1(3): 233-237.

Vogel, S. 1999. *The skinny on Fat.* W.H. Freeman and Co., New York.

Volek, J.S., Phinney, S.D., Forsythe, C.E. et al. 2009. Carbohydrate restriction has a more favorable impact on the metabolic syndrome than a low fat diet. *Lipids* 44(4): 297-309.

Walker, L.C. 1998. *The North American forests: Geography, ecology and silviculture.* CRC Press, London.

Ward, B. 1976. *The home of man.* W.W. Norton & Company, New York.

Warman, A. 2003. *Corn & capitalism: How a botanical bastard grew to global dominance.* Trans. From Spanish by Westrate, N.L. Univ. of North Carolina Press, Chapel Hill, NC.

Westman, E.C., Volek, J.S. and Feinman, R.C. 2006. Carbohydrate restriction is effective in improving atherogenic dyslipidemia even in the absence of weight loss. *American Journal of Clinical Nutrition* 84(6): 1549.

Westman, E.C., Yancy, W.S., Edman, J.S. et al. 2002. Effect of 6-month adherence to a very low carbohydrate program. *American Journal of Medicine* 113(1): 30-36.

Wiley, S.T. 1883. *History of Monongahela county, West Virginia, from its early settlements to the present time; with numerous biographical and family sketches.* Preston Publishing Company, Kingwood, WV.

White, W.J. 2000. *An unsung hero: The farm tractor's contribution to twentieth-century United States economic growth.* Ph.D. Dissertation. Ohio State University, Columbus, OH.

Williams, W.W. 1996 (December). From Asia's good earth: rice, society & science. *Hemispheres*: 80-88.

Wolf, E.R. 1966. *Peasants.* Prentice-Hall, Englewood Cliffs, NJ.

Yanovski, S.Z., Reynolds, J.C., Boyle, A.J. and Yanovski, J.A. 1997. Resting metabolic rate in African-American and Caucasian girls. *Obesity Research* 5(4): 321-325.

Yerushalmy, J. and Hilleboe, H.E. 1957. Fat in the diet and mortality from heart disease; a methodologic note. *New York State Journal of Medicine* 57(14): 2343-2357.

Zhang, Y., Proenca, R., Maffei, M. et al. 1994. Positional cloning of the mouse obese gene and its human homologue. *Nature* 372: 425-432.

Index

C

D

E

F

G

H

I

T

U

V

W

Made in the USA
Monee, IL
12 November 2026

8e6a3767-fb8c-4edb-a3a7-14b701ffc8ccR01